Fibre production in South American camelids and other fibre animals

Fibre production

in South American camelids and other fibre animals

edited by:

Mª Ángeles Pérez-Cabal

Juan Pablo Gutiérrez

Isabel Cervantes

Mª Jesús Alcalde

Wageningen Academic Publishers

ISBN: 978-90-8686-172-9
e-ISBN: 978-90-8686-727-1
DOI: 10.3921/978-90-8686-727-1

Cover photo: Blanca Nieto at
Pacomarca ranch

First published, 2011

Wageningen Academic Publishers
P.O. Box 220
6700 AE Wageningen
The Netherlands
www.WageningenAcademic.com
copyright@WageningenAcademic.com

Preface

The main objective of these European meetings is to promote the exchange of scientific information among international groups dedicated to the production of animals producing textile fibre. They are organised under the umbrella of the European Association of Animal Production (EAAP) and claims for relaunching the meeting activities of the Animal Fibre Working Group organised within the EAAP. The previous European Symposium was held in Göttingen (Germany) in 2006 and our main aim was to set a 3-year frequency for the following meetings to continue with the debates about South American Camelids and Fibre Animals.

The 5th European Symposium on South American Camelids and First European Meeting on Fibre Animals was jointly organised by the Complutense University of Madrid (UCM) and the University of Seville (US) in Seville (Spain). Dr. Juan Pablo Gutiérrez (UCM) was the Chairman of the Organising Committee which was formed by M[a] Jesús Alcalde (US), M[a] Ángeles Pérez-Cabal (UCM), Ester Bartolomé (US), Rocío Álvarez (US), Mercedes Valera (US) e Isabel Cervantes (UCM). We wish to thank especially Andrés Pérez and Meritxell Justicia, students from both universities, who helped us during the Symposium.

The Symposium was held at the 'Diputación of Seville'. We thank the administration of the Diputación for providing the lecture and posters halls and all the staff for their help.

We are grateful to the 'Ministerio de Ciencia e Innovación' (Spanish Ministry of Science and Technology) for supporting the attendance of the invited speakers. ETSIA and Vicerrectorado de Relaciones Institucionales of University of Seville, Barclays Bank, Diputación of Seville, Turismo of Seville Convention Bureau, Ministerio de Medio Ambiente y Medio Rural y Marino, and Consejería de Agricultura y Pesca of Junta de Andalucía funded the organisation of the Symposium. We would also like to acknowledge the 'Instituto Nacional de Investigación y Tecnología Agraria y Alimentaria' (INIA, Spanish Institute of Agricultural Technology and Research) who funded these proceedings (AC2010-000002-00-00), that were printed from the manuscripts supplied by the authors. We would like to thank the editors of Wageningen Academic Publishers. Here are the studies presented in Seville that the contributors wanted to be published in these proceedings.

M[a] Ángeles Pérez-Cabal, Juan Pablo Gutiérrez, Isabel Cervantes, and M[a] Jesús Alcalde
The editors

Table of contents

Nutrition and reproduction

Management

Health

Abstracts

Round tables

Keyword index

Setting the scene: animal fibre

Genetic variability of fleece shedding in the Martinik hair, Romane sheep breeds and their crossbreds

D. Allain[1], B. Pena[1], D. Foulquié[2], Y. Bourdillon[3] and D. François[1]
[1]*INRA, UR 631, Amélioration Génétique des Animaux, F-31326 Castanet Tolosan, France; daniel.allain@toulouse.inra.fr*
[2]*INRA, UE 321, Domaine de la Fage, Saint Jean et Saint Paul, 12250 Roquefort, France*
[3]*INRA, UE 332, Domaine de Bourges-la Sapinière, 18390 Osmoy, France*

Abstract

From the ancestral pelage of the domestic sheep which looks like the protective double coat of the wild sheep, changes following domestication and the development of textile industry resulted in the development of the modern woolled sheep with a large decrease in the tendency to moult and extension of wool fibre at the expense of coarse hair leading to a single coat where all fibres are similar in dimension and grow permanently. Wool production is still the main purpose of sheep farming in the southern hemisphere, but in Europe wool production is often unprofitable and there is an interest in the use of breeds that have no wool or shed their wool due to the relative value of meat and wool, and increasing shearing costs. In the present paper investigations were made to evaluate genetic variability of moulting or fleece shedding including QTL detection in the French Romane, the Martinik Hair, a hairy sheep without wool close to Barbados-Black-Belly and a Martinik Hair-Romane backcross population. In the Romane breed, 43.2% of adult ewes shed at least part of their fleece once a year during spring. Heritability estimate of moulting aptitude is high (0.46) but prevalence of total fleece shedding is low (1.5% of ewes shed all their fleece). In the Martinik Hair-Romane backcross population 72% of animals shed their fleece. It was proposed to introgress gene pool from the Martinik Hair in the Romane breed through an experimental design combining QTL detection and introgression of QTL for fleece shedding in the Romane breed.

Keywords: sheep, fleece shedding, genetic variability, QTL detection, gene introgression

Introduction

The wild ancestor of sheep was a hair sheep which evolved into modern woolled breeds through domestication, migration of people from the Middle East and artificial selection. The first important migration was of hair sheep; the present-day tropical hair sheep and some other hair breeds (Wiltshire Horn, Soay, …) are descendants of these sheep. Later, migrations of people from the Middle East outwards to south-east Asia, Africa and Europe were accompanied by domestic animals, including the more recently developed woolled sheep. In time, these sheep displaced hair sheep everywhere, except the humid tropics (Zygoyiannis, 2006).

The needs of humans for clothing, and of the sheep for protection from cold, were undoubtedly responsible for the evolution of the many woolled breeds and their dominance in both Europe and the arid regions of Asia and Africa. In particular, the Romans introduced a white woolled sheep into Western Europe and this was a major contributor to the British breeds of sheep, and probably also fine woolled sheep. The development of true fine wool began in the Middle East probably soon after 1000 BC (Ryder, 1969). But the emergence of the Merino as a distinct breed occurred in the late Middle Ages in Spain (Ryder, 1984). Wool production based on the Merino-breed was then developed in the New World and southern hemisphere from the 19th century.

Along with domestication and selection process in response to the human needs for clothing and development of the textile industry, the coat of the wild sheep evolved into the merino fleece. The coat of the wild sheep and derived hair sheep comprises a hairy outer coat (composed of kemp) and a woolly undercoat. This coat has a simple cycle of active growth from spring to autumn followed by inactivity in winter, which is annually driven by seasonal changes in daylight. The whole coat moults every spring with replacement of outer coat only from spring to summer and then replacement of the undercoat from summer to autumn. Changes following domestication and selection have resulted in (1) an increasing period of hair growth leading to a growth phase longer than a year and subsequently a decrease in the tendency to moult, and (2) the development of the wool at the expense of the hair and kemp. As a consequence for textile use, the lengthening of the hair growth phase provides the necessary fibre length for spinning while extension of the wool undercoat leads to the modern fine woolled sheep owning a single coat where all fibres are similar in dimensions and grow permanently (Rougeot, 1982).

Today wool production remain the main purpose of sheep farming in the southern hemisphere, mainly Australia, New Zealand, Argentina, Uruguay, South Africa and Asia. In contrast, in Europe wool production is now generally unprofitable and may indeed be undesirable compared to meat or milk production. This is mainly because the income from wool is less than the costs associated with shearing and the removal of soiled wool ('dagging') from lambs prior to slaughter or to prevent fly strike (Vipond, 2006). Consequently with increasing shearing costs, there is a new interest in Europe for the use of breeds that have no wool or shed their wool. Some farmers in the UK or Germany are experimenting by crossing their sheep with other breeds that either shed their wool annually, or are recognised 'hair' sheep breeds, such as the Wiltshire Horn, the Dorper, Khatardin and Barbados Blackbelly, with varying degrees of success (Conington, 2010). To date no similar experiments have been undertaken in France. The present paper describes experimental designs with first results aiming (1) to investigate genetic variability including QTL detection of fleece shedding aptitude in the Martinik Hair sheep breed, the French Romane breed and their crossbred and (2) to evaluate genetic strategies for gene introgression of fleece shedding in the Romane breed in France.

Fleece shedding aptitude in the Romane breed

Material and methods

The Romane breed is a composite line of 50% Romanov (prolific breed) and 50% Berrichon du Cher (meat breed), considered as a new breed at the 4th intercross generation (Ricordeau *et al.*, 1992). Selection objectives of this breed are mostly maternal traits (prolificacy, growth of the lambs between 10 and 30 days) and meat production traits. This breed is known to have a high productive potential but shows a large variability in its fleece type due to the variability of coat structure in parental lines. Romanov breed has a coarse fleece with some fibre shedding (Bykova, 1973) while Berrichon du Cher derived from Merino by crossbreeding during the 19th century has a fine medium crossbred fleece. The Romane breed, shows a large variability in its fleece type with a high variable birthcoat type in lambs related to lamb survival and a kemp wool fleece of a poor quality with some aptitude for fleece shedding (Allain *et al.*, 2010b).

General experimental conditions

A flock of 350 ewes of the Romane breed has been used as experimental support since 2000 at the INRA farm of La Fage on the Causses-du-Larzac, a calcareous plateau at an altitude of 800 m in the south of France. This flock was maintained in permanent outdoors conditions on rangelands with arid conditions despite an abundant annual rainfall of 1000 mm, due to an important permeability of the soil, which feeds deep subterranean rivers. Seasons are highly contrasted with cold winter due to the altitude, hot summer due to the southern latitude and intermediate seasons showing high variations in temperature, wind and rainfall. The breeding system thus described (Bouix *et al.*, 2002) is characterised by a short lambing period in outdoors conditions from end-March to mid-April. Climate conditions in that period can vary abruptly from severe conditions of low temperature, wind and rain or snow to warm and sunny conditions. The annual shearing time for ewes is the beginning of July, just after weaning the lambs. A total of 3,051 observations on fleece shedding aptitude were made once a year from the age of 14 months on 1,359 ewes from 79 sires just prior to the annual shearing at the end of June.

Fleece shedding measurement

Fleece shedding or moulting aptitude was determined in adult animals once a year at the end of June just before shearing time. In a first step, at farm level, the part of the animal body area denuded or without wool was drawn on a standard sheep profile. Then this figure was analysed using image software to estimate the extent of moulting or fleece shedding as a percentage of the total body area.

Data analysis

Data were analysed by ANOVA with the GLM procedure of the SAS package. In the analysis model, the considered fixed effects were year of observation (from 2000 to 2009), age (4 levels from 1 year to 4 years old and more), litter size (4 levels from none to 3) and number of suckling lambs. Genetic parameters of fleece shedding aptitude in the adult Romane ewe were estimated using a TM (Threshold Model) programme[1] using a Bayesian analysis and performing numerical integration through the Gibbs sampler. Fleece shedding rate was considered as a continuous trait[2].

Results and discussion

Fleece shedding rate and moulting pattern

Annual fleece shedding aptitude in the 1,359 adult Romane ewes observed once a year (3,051 records) is reported in Figure 1. Mean fleece shedding rate is 17.0% with 43.2% and 1.4% of adult ewes which shed at least part of and all their fleece once a year during spring, respectively. As a general pattern of moulting over body regions, wool shedding was first observed from neck and belly, then moved progressively in a dorso-posterior direction to end on the rump. The Romane breed being a composite line of 50% Romanov and 50% Berrichon du cher, gene pool involved in fleece shedding aptitude is probably originated from the Romanov breed where fibre shedding has been reported (Bykova, 1973) while no fleece shedding has ever been reported in the Berrichon du Cher breed.

Non-genetic effects and heritability of fleece shedding in the adult Romane ewe

Age of the ewe, reproductive activity and year of observation have a highly significant effect on fleece shedding aptitude. Young ewes of 1-year old shed significantly less ($P<0.001$) their fleece than older ewes. Increase of wool shedding with age has also been reported in Wiltshire-Merino crosses (Rathie *et al.*, 1994). At one year of age hog ewes shed less ($P<0.05$) than breeding ewes. Such a difference in fleece shedding aptitude in young ewes can be due either to gestation and

---

[1] Available at http://snp.toulouse.inra.fr/~legarra or upon request to the author at: andres.legarra@toulouse.inra.fr.

[2] Authors thank Andres Legarra for help in using the TM programme.

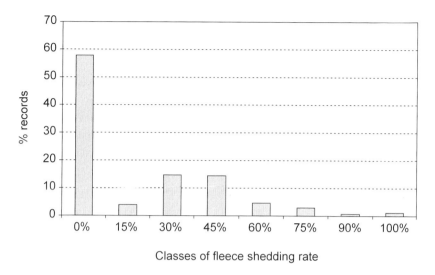

Figure 1. Fleece shedding in the Romane breed categorised by percentage classes of body surface where fleece is shed in adult ewes (3,051 records from 1,359 ewes).

reproductive activity or to body growth development at the age of 7 months as the decision for the age of first mating (7 or 19 months) is conditioned by a minimum live weight of 35 kg at 7 months of age. But in breeding ewes, whatever the age, there was no effect of the litter size on fleece shedding.

High heritability (0.46±0.06) estimate of fleece shedding aptitude was observed in the Romane breed (Figure 1). Genetic variability of fleece shedding aptitude has never previously been reported in a woolled sheep. It offers opportunities to implement selection leading to a moulting sheep requiring no shearing and thus creation of genetic progress by improving adaptive traits in a breed with a high production potential (Francois *et al.*, 2010).

In addition to fleece shedding, the Romane breed shows another fleece adaptive trait to be selected in extensive management under permanent exposure outdoors and harsh environments. At birth lambs exhibit a high variability of coat type from a long hairy coat to a very short woolly coat and lamb survival rate is related to birthcoat type (Allain *et al.*, 2009). Lambs bearing a long hairy double coat have a higher survival rate than lambs with a short woolly coat. This difference in lamb survival (4 points; 93% vs. 89%) could be explained by a lesser thermal protection with higher heat losses and a lesser resistance to cold and wet weather conditions in lambs bearing a short woolly coat (Allain *et al.*, 2010a). High heritability estimates of both birthcoat type (0.58) and fleece shedding (0.47) were observed but there were no genetic correlations between these two fleece adaptive traits (Allain *et al.*, 2010b) suggesting that different genes are controlling determinism of fleece shedding and birthcoat hairiness.

Fleece shedding aptitude in Martinik hair and Martinik hair-Romane backcross

Material and methods

General experimental conditions

A flock of 50 Martinik Hair ewes and a flock of 800 Romane ewes breeds were used as experimental support at the INRA farm of La Sapinière (Domaine experimental de Bourges). Animals are raised in a barn with open shed. Depending on grazing availabilities and season, animals are fed indoors and go to pasture during the day from spring to autumn. As concerns the reproduction cycle, Romane ewes were mated for the first time at the age of 7 months in July, at the age of 16 months in April for the second time and then once a year in April. Martinik Hair ewes were mated once a year in July.

Martinik Hair sheep

The Martinik Hair sheep is a typical double coat hair sheep with no wool, originated from the French West Indies but related (origins and history) to the other hair sheep populations present in the Caribbean Islands and Central America (Barbados Blackbelly, Pelibuey, West African Dwarf) (Leimbacher *et al.*, 2010). The Martinik Hair sheep has a high production potential (prolificacy, low susceptibility to endoparasites) and was introduced in the form of embryos in France during the nineties. A small flock of 50 ewes was created at the INRA farm of La Sapinière (Domaine de Bourges) for experimentation. As in tropical conditions, this hair sheep also moults and renews its whole fleece annually from spring to summer in temperate conditions.

Creation of a backcross Martinik Hair-Romane sheep population

In 2001, 4 Martinik Hair rams were mated with 100 Romane ewes to produce F1 rams. 4 sons of the different Martinik Hair sires were mated with 180 Romane ewes to produce 228 backcrossed animals born in December 2006. All these backcrossed animals were measured for fleece shedding aptitude in late spring at 7 months. Thereafter within this population, 83 animals were also measured for fleece shedding aptitude once a year in late spring for 2 years, i.e. at 19 and 31 months of age.

An inter-backcross population was also created in 2008. 8 backcross sires were mated to backcross ewes to create 102 individuals of the inter-backcross population which were also measured for fleece shedding aptitude at 7 months of age in late spring. Fleece shedding aptitude was measured in a similar way to that previously for Romane ewes.

Results and discussion

Annual fleece shedding aptitude in Martinik hair - Romane backcross animals

Figure 2 shows fleece shedding aptitude at 7 months of age in the 228 backcross animals. 71.9% of individuals shed at least 30% of their fleece during spring and 2 animals shed all their fleece with a mean fleece shedding rate of 36.4%. Non-genetic effects such as sex, age of dam or litter size were tested but any significant effect was not observed. In contrast, a highly significant sire effect was observed. Within the sire family (44 to 75 halfsibs/sire family), the mean proportion of fleece shedding varies from 18.6 to 47.0% between the 4 family sires studied. The general pattern of wool shedding over body regions is similar to that observed in the pure Romane breed.

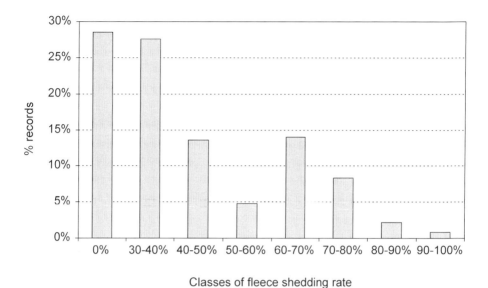

Figure 2. Fleece shedding categorised by percentage classes of body surface where fleece is shed at 7 months of age in 228 Martinik Hair-Romane backcross animals.

Fleece shedding was then observed in 83 animals of both sexes over 3 consecutives years. Data were analysed using GLM and then Varcomp SAS procedures to determine variability and animal variance components of fleece shedding in the 83 animals observed 3 times at yearly intervals. Age and sex-age interaction effects were observed. Fleece shedding increases with age from a rate of 30.3% at both 7 and 19 months of age to 45.0% at 31 months of age as previously observed in the Romane breed and Wiltshire-merino crosses (Rathie *et al.*, 1994). There was no sex effect at 7 months of age but thereafter highly significant differences were observed according to sex at 19 and 31 months of age. Compared to males, a higher and a lower fleece shedding were observed in females at 19 and 31 months of age, respectively. As a possible explanation, sex hormones and/or genes on the sexual chromosome could be involved in determining fleece shedding. Repeatability estimated using the SAS VARCOMP procedure was 0.59.

Fleece shedding in the inter-backcross population of Martinik Hair-Romane animals

Figure 3 shows the fleece shedding rate in the 102 animals of the inter-backcross population. 60.8% of individuals shed at least 30% of their fleece during spring and 4 animals shed all their fleece. Compared to backcross animals, a smaller proportion of animals shed part of their fleece but a larger part shed all their fleece.

In agreement with Slee (1959), who studied different crosses between the Wiltshire Horn and the Scottish Blackface, at least two components of the shedding process can be distinguished: (1) the ability to shed and, (2) the extent of shedding or proportion of the body area defleeced. In the present work, by comparing backcross population to parental pure breed lines, there is evidence that genetic determinism of the fleece shedding extent is a polygenic effect but the existence of major genes for the ability to shed cannot be excluded. The mean fleece shedding rate observed in backcross animals (owning ¾ of the Romane gene pool) is close to expected value estimate of the shedding rate in parental lines (36.4%, 100% and 17% in Martinik Hair-Romane backcross, Martinik Hair and Romane respectively), but the proportion of animals that shed at least a

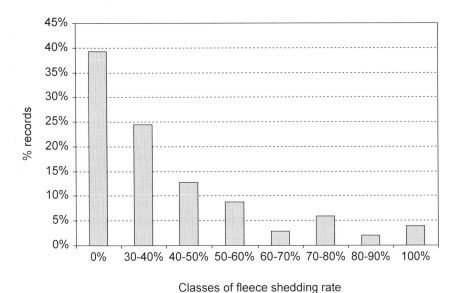

Figure 3. Fleece shedding categorised by percentage classes of body surface where fleece is shed at 7 months of age in 102 inter-backcross Martinik Hair-Romane animals.

part of their fleece in backcross animals is intermediate between the two parental lines (100%, 43.2% and 71.9% in Martinik Hair, Romane and backcross animals, respectively) suggesting a deviation from additivity with major gene and/or dominant effects of the ability to shed.

QTL detection for fleece shedding

In parallel to these investigations, two experimental designs for QTL detection of fleece shedding aptitude through whole genome scans were initiated from 2004. The first one using the Romane breed population at INRA La Fage experimental farm comprises 10 sire families of 100 halfsibs/ sire for QTL detection of birthcoat type, fleece shedding and maternal behaviour traits (Boissy et al., 2007). Within this experimental design, a total of 600 adult ewes from 10 sire families were observed once a year for fleece shedding aptitude just prior to the annual shearing at the end of June.

The second one is based on the Martinik Hair-Romane backcross population at La Sapinière Inra experimental farm and comprised 4 sire families of 60 halfsibs/sire for QTL detection of resistance to parasitism and fleece shedding (Moreno et al., 2006). Animals of these QTL experimental designs have been genotyped with the OvineSNP50 bead chip (Moreno et al., 2009), and analyses are in progress to detect QTL and then to fine-map relevant genes controlling fleece shedding aptitude in these 2 population resources as well as birthcoat type in the Romane breed.

Genetic strategies for gene introgression of fleece shedding gene in the Romane breed

Under European conditions where the value of wool is often lower than shearing costs and other costs associated with managing wool, selection for fleece shedding in order to suppress shearing represents an alternative opportunity for improving the profitability of sheep farming. However, as for any other adaptive trait, selection strategies for fleece shedding in the Romane breed could be selecting together with production traits according to two different strategies.

The first strategy could be to select within the Romane breed as according to our results, a genetic variability with a high heritability estimate of fleece shedding aptitude was observed offering opportunities for within-breed selection leading to a moulting sheep requiring no shearing. However, it is not easy to measure this trait and prevalence of total fleece shedding is low. An alternative method could be to introgress gene pool controlling fleece shedding from the Martinik Hair, a 'hair sheep'.

In order to valorise results observed up to now on fleece shedding in both the Romane breed and the Martinik Hair-Romane backcross population, an experimental design aimed at introgressing gene pool for fleece shedding from the Martinik hair breed into the Romane breed through 4 successive backcrossing generations was initiated in 2009. The experimental design is similar to those combining detection and introgression QTL in outbred populations recently proposed by Yazdi *et al*. (2008, 2010) as an efficient method for saving time and cost in an introgression process. For instance, 6 Martinik Hair-Romane backcross rams with total or near total fleece shedding phenotype from the QTL experimental design described above, were mated with 120 hog Romane ewes expressing partial fleece shedding at 7 months of age in late spring. Measurements made with 360 hog ewes showed that about 40% of hog Romane ewes shed part of their fleece at 7 months of age. Similar mating schemes will also be used for producing the last backcrossing generation. The use of molecular information with QTL introgression will be very helpful in the near future, as QTL detection through a whole genome scan with the OvineSNP50 bead chip is in progress in both the Romane breed and the Martinik Hair-Romane backcross population.

Conclusions

Genetic variability of fleece shedding was investigated in the Romane breed and a Romane-Martinik hair backcross population. The high heritability estimate of fleece shedding observed in the Romane breed offers opportunities to implement selection, leading to a moulting sheep requiring no shearing as a new adaptive trait, together with production traits in this breed known to have a high productive potential.

Despite high genetic variability of fleece shedding in the Romane breed, it was proposed to increase efficiency by introgressing a moulting gene pool with the use of molecular markers from the Martinik Hair breed, a hair sheep with no wool, due to low prevalence of total fleece shedding in the Romane breed. The experimental design scheme will combine detection and introgression of QTL for fleece shedding in the Romane breed.

References

Allain D., D. Foulquié, D. Francois, P. Autran, B. Bibé and J. Bouix, 2009. The birthcoat type: an important component of lamb survival in the French Romane breed raised under permanent exposure outdoors. In: Proceedings "EAAP2009", Barcelona, 24-27 August, Abstract no. 4195.

Allain D., D. Foulquié, D. Francois, B. Pena, P. Autran, B. Bibé and J. Bouix, 2010a. Birthcoat type as an important component of lamb survival for extensively managed sheep in the Romane breed: genetic variability and QTL detection experimental design. In: D. Allain (ed.), Proceedings "8th World Merino Conference", Rambouillet, France, 3-5 May, 2010, com no 5-16, pp. 5.

Allain D., D. Foulquié, D. Francois, B. Pena, P. Autran, B. Bibé and J. Bouix, 2010b. Birthcoat type and fleece shedding as adaptative traits for extensively managed sheep: genetic variability and QTL detection experimental design. In: Proceedings "9th World Congress on Genetics Applied to Livestock Production", Leipzig, Germany, 1-6 August, 2010, com no 348, pp. 4.

Boissy A., S. Ligout, D. Foulquié, A. Gautier, C. Moreno, E. Delval, D. François and J. Bouix, 2007. Analyse génétique de la réactivité comportementale chez les ovins: pour une stratégie combinant bien être et production (Genetics of behavioural reactivity in sheep: a strategy for combining animal welfare and efficiency of production), Rencontres Recherches Ruminants, 14: 301-304.

Bouix J., M. Jacquin, D. Foulquié, P. Autran, P. Guillouet, D. Hubert and B. Bibe, 2002. Genetic effects on sheep production in harsh environment and extensive management system. In: Proceedings "7th World Congress on Genetics Applied to Livestock Production", Montpellier, France, August 19-23, 2002, Communication N° 02-02, 441-118.

Bykova G.L., 1973. The effect of season on the fleece of Romanov sheep of various classes, Trudy Yaroslavskogo Nauchno-Issledovatel'skogo Instituta Zhivotnovodstva i Kormoproizvodstva, (3) CABI:19740108295

Conington J., 2010. Sheep breeding, production and conservation grazing in the UK- An example from Northern Europe. In: D. Allain (ed.), Proceedings "8th World Merino Conference", Rambouillet, France, 3-5 May, Com no 3-01.

François D., A. Boissy, P. Jacquiet, D. Allain, B. Bibe, R. Rupp, C. Moreno, L. Bodin, D. Hazard and J. Bouix, 2010. Genetics of adaptation traits for harsh environment in sheep. In: Proceedings "9th World Congress on Genetics applied to Livestock Production", Leipzig, Germany, 1-6 August, com no 453, pp. 4.

Leimbacher F., G. Alexandre, M. Mathieu, M. Naves and N. Mandonnet, 2010. The Martinik Hair sheep: a high potential breed to produce mutton in the tropics. In: D. Allain (ed.), Proceedings "8th World Merino Conference", Rambouillet, France, 3-5 May, Com no 5-04.

Moreno C., L. Gruner, A. Scala, L. Mura, L. Schibler, Y. Amigues, T. Sechi, P. Jacquiet, D. François, S. Sechi, A. Roig, S. Casu, F. Barillet, C. Brunel, J. Bouix, A. Carta and R. Rupp, 2006. QTL resistance to internal parasites in two designs based on natural and experimental conditions or infections. In: D. Allain (ed.), Proceedings "8th World congress on genetics applied to livestock production", Belo Horizonte, MG, Brazil, Com no 15-05 (CD-Rom: Article_15_645-1640.pdf).

Moreno C., B. Servin, T. Faraut, C. Klopp, R. Rupp, P. Mulsant, C. Robert-Granier, F.Barillet, C. Delmas, J. Bouix, D. Francois, D. Allain, E. Manfredi, L. Bodin, J.M. Elsen, D. Robelin, B. Mangin, M.R. Aurel, F. Bouvier, D. Calavas, M. Sancristobal, P. Jacquiet, G. Foucras, A. Boissy and A. Legarra, 2009. SheepSNPQTL: Utilisation d'une puce 60 000 SNP pour cartographier finement des QTL affectant des caractères de production, de résistance aux maladies et de comportement chez les ovins (Use of a 60K SNP array for QTL detection and fine mapping production, resistance to disease and behaviour traits in sheep). Rencontres Recherches Ruminants, 16: 420.

Rathie K.A., M.L. Tierney and J.C. Mulder, 1994. Assessing Wiltshire Horn - Merino crosses 1. Wool shedding, blowfy strike and wool production traits. Australian Journal of Experimental Agriculture, 34: 717-728.

Ricordeau G., L. Tchamitchian, J.C. Brunel, T.C. Nguyen and D. François, 1992. La race ovine INRA 401: un exemple de souche synthétique (The INRA401 sheep breed: an example of a composite line), Productions Animales, hors série: "Eléments de génétique quantitative": 255-262.

Rougeot J., 1982. Evolution de la toison en relation avec les caractéristiques textiles (Changes in coat structure and composition in relation to textile characteristics). Bulletin Scientifique ITF, 11 (41): 41-52.

Ryder M.L., 1969. Changes in the fleece of sheep following domestication (with a note on the coat of cattle). In: P.J. Ucko and G.W. Dimbleby (eds.), The domestication and exploitation of plants and animals. Aldine Publishing, Chicago, IL, USA, pp. 495-521.

Ryder M.L., 1984. Medieval sheep and wool types, The Agricultural History Review, 32: 14-28.

Slee J., 1959. Fleece shedding, staple length and fleece weight in experimental Wiltshire Horn-Scottish Blackface sheep crosses. Journal of Agricultural Science, 53: 209-223.

Vipond J., 2006. Easicare sheep systems. SAC Select Services, Penicuik, Scotland.

Yazdi M., A. Sonesson, J. Woolliams and T. Meuwissen, 2008. Combined detection and introgression of QTL quantitative trait loci underlying desirable traits. Journal of Animal Science, 86: 1089-1095.

Yazdi M.H., A. Sonesson, J. Woolliams and T. Meuwissen, 2010. Combined detection and introgression of QTL in outbred populations, Genetics Selection Evolution, 42: 16.

Zygoyiannis D., 2006. Sheep production in the world and in Greece. Small Ruminant Research, 62: 143-147.

Suri/Huacaya phenotype inheritance in alpaca (*Vicugna pacos*)

C. Renieri[1], A. Valbonesi[1], M. Antonini[1], V. La Manna[1], T. Huanca[2], N. Apaza[2], S. Presciuttini[3] and M. Asparrin[4]
[1]*School of Environmental Sciences, University of Camerino, Via Gentile III da Varano s.n.c., 62032 Camerino, Italy; carlo.renieri@unicam.it*
[2]*INIA, ILLPA Puno, Rinconada Salcedo, Puno, Peru*
[3]*Department of Physiological Sciences, University of Pisa, Via San Zeno 31, 56123 Pisa, Italy*
[3]*Michell y CIA S.A., Fundo Mallkini, Peru*

Abstract

The Suri/Huacaya phenotype inheritance in alpaca was tested on two independent Peruvian sources of records: the Registry of Mallkini farm (588 offspring by Suri sire × Suri dam from 62 paternal half sib families, and 2,126 offspring by Huacaya sire × Huacaya dam from 177 paternal half sib families) and the results of the Quimsachata INIA ILPA Puno experimental trial (two reciprocal experimental test-crosses, involving a total of 17 unrelated males and 149 unrelated females). The data support a genetic model in which two linked loci must simultaneously be homozygous for recessive alleles in order to produce the Huacaya phenotype. The estimated recombination rate between these loci was 0.099 (95% C.L. = 0.029-0.204). The birth of 3 Suri offspring from Huacaya x Huacaya mating is explained by a new dominant mutation on some germinal lines of Huacaya animals. The direct mutation rate can be estimated at 0.0014.

Keywords: alpaca, Suri, Huacaya, segregation analysis, linkage

Introduction

This paper summarises the results of the research on Suri/Huacaya phenotypes inheritance carried out at the Italian University of Camerino, School of Environmental Sciences in collaboration with the Peruvian private Michel Group and the Peruvian National Institute for Agronomic Innovation (INIA, ILLPA Puno). The results are published in Renieri *et al*. (2009) and Presciuttini *et al*. (2010).

The inheritance of Suri/Huacaya phenotypes: a review

The inheritance mode of the Suri/Huacaya phenotype is unclear. A hypothesis of Suri recessivity has been abandoned in favour of a more widely accepted simple Suri dominance. Velasco (1980) proposed a single locus model with a dominant allele for Suri. Similarly, Ponzoni *et al*. (1997) suggested a model with a single gene with two alleles, though they also noted that the data were compatible with a model in which the trait was controlled by a group of closely linked alleles (haplotype). Baychelier (2000) tested three inheritance hypotheses using the records of the Australian Alpaca Association Herd Book: the model of a single gene with three alleles can be rejected, whereas the two-gene model is more suitable than the one-locus model. Finally, Sponenberg (2010) proposed a single autosomal dominant gene, with an additional genetic mechanism that can suppress the Suri phenotype in some animals. It should be noted that 4 Suri offspring have been observed among the offspring of 19,637 Huacaya × Huacaya crosses.

Experimental designs

Two sets of records were analysed. The first source is the Registry record of the Mallkini farm, the private farm of Michell Group. In this case, 588 (291 females and 297 males) offspring by Suri sire × Suri dam from 62 paternal half sib families, and 2,126 (1,009 females and 1,117 males) offspring by Huacaya sire × Huacaya dam from 177 paternal halfsib families, born over 4 years (2004, 2005, 2006, and 2007), were analysed.

The second set of data was recorded in the crosses carried out at the experimental station of the INIA located in Quimsachata, Peru. The trial involved a total of 17 unrelated males and 149 unrelated females. Two reciprocal experimental test-crosses were carried out: Suri males × Huacaya females and Huacaya males × Suri females: 64 total Suri animals and 102 total Huacaya animals.

Results and discussion

Single gene hypothesis

The single gene hypothesis was tested in both data sets. In the Mallkini records, assuming that the percentage of the two female genotypes is the same (50%), the expected segregation 7 Suri: 1 Huacaya individual was tested. The results are presented in Table 1.

With the exception of the offspring of male 0-258, in all the remaining 18 segregations the differences between the observed and the expected frequencies are not statistically significant ($0.057 \leq P \leq 0.924$). Similar results were obtained from aggregated data analysis; the heterogeneity G-test indicates that the total G (G_T=20.276, P=0.378), as well as its components (G_P=0.347, P=0.556; G_H=19.929, P=0.33).

Considering first the Suri males in the Quimsachata experimental trial (Table 2), for the eight segregating animals it was possible to estimate the segregation ratio R of the Huacaya phenotype among their offspring assuming a truncated binomial distribution. With this model the maximum likelihood estimate of R was 0.290, with 95% C.L.=0.184-0.409. Clearly, a value of R=0.5, which is expected for a testcross of a recessive single-locus model can be rejected.

With regard to the reciprocal cross a similar analysis cannot be carried out. When one offspring only can be obtained from a Suri animal (the female in this case), the observed segregation ratio for a number of animals of unknown genotype is a function of two parameters: (1) the probability R that a segregating animal generates a Huacaya offspring (e.g. the Mendelian transmission probability), and (2) the probability H that an animal is segregating. A practical approach could be to estimate the value that each parameter assumes, once the other parameter is fixed on a value representing a reasonable choice. Assuming the segregation ratio R=0.5 (the probability that a Suri female mated to a Huacaya male produces a Huacaya offspring for a simple recessive genetic model), the maximum likelihood estimate of H is 0.582. Conversely, if we assume that H among the Suri females has the same value as that observed among Suri males (= 8/9, or 0.88), the maximum likelihood estimate of R is 0.331. The similarity of this last value with that estimated for the crosses of Suri males is certainly interesting. The results are for the rejection of the single-locus recessive model for the Huacaya phenotype.

Table 1. Observed and expected frequencies of Suri × Suri segregating families in a tested 7:1 hypothesis.

Males	Offspring	Observed frequency		Expected frequency[1]		G_{adj}[2]	P
		Suri	Huacaya	Suri	Huacaya		
190	32	31	1	27.9	4.1	3.630	0.057
239	18	16	2	15.5	2.5	0.111	0.739
65691	27	26	1	23.5	3.5	2.697	0.101
00-024	27	23	4	23.5	3.5	0.089	0.765
00-034	30	27	3	26.2	3.8	0.215	0.643
0-176	27	26	1	23.5	3.5	2.697	0.101
0-258	19	13	6	16.4	2.6	4.053	0.044
02-848	34	29	5	29.7	4.3	0.126	0.722
1-1-245	6	5	1	4.6	1.4	0.133	0.716
1-2-240	9	7	2	7.4	1.6	0.109	0.741
1-3-214	8	7	1	6.5	1.5	0.246	0.620
1-3-310	12	10	2	10.1	1.9	0.009	0.924
1-3-329	10	8	2	8.3	1.7	0.063	0.802
1-3-333	10	9	1	8.3	1.7	0.392	0.531
2-1-376	19	16	3	16.4	2.6	0.076	0.783
2-3-335	5	4	1	3.7	1.3	0.089	0.766
2-3-342	5	3	2	3.7	1.3	0.489	0.484
9-163	20	19	1	17.3	2.7	1.553	0.213
9-510	44	34	10	38.5	5.5	3.471	0.062
Pooled	362	313	49	316.8	45.3	0.347	0.556

[1] According to the formula suggested by Andresen (1974).
[2] Adjusted G-values according to William's correction.

Double genes, independent segregation (epistasy)

Three independent two loci-two phenotype hypothesis were tested through the Mallkini records. In the 15:1 segregation hypothesis for duplicate Suri dominant gene action (Table 3), two progenies (the offspring of males 0-258 and 9-510), both with an excess of Huacaya and a low number of Suri, showed significantly different results than expected (P=0.05 and 0.001, respectively). The resulting heterogeneity G is clearly not significant (G_H=4.56; P=0.999), indicating that most of the deviations from expectation of the progenies are in the same direction and are not significantly different from each other. The pooled G has a highly significant value (G_P=25.051; P<0.001) and the total G shows a borderline value (G_T=29.65; P=0.056).

In the 13:3 segregation hypothesis (Table 4) for dominant suppression epistasis (the dominant gene for Suri is supposed to mask the genes at the Huacaya locus) three progenies (the offspring of males 190, 65691, and 0-176), each with an excess of Suri and a low number of Huacaya, displayed significantly different results than expected (P=0.007, 0.018, and 0.018, respectively). Although none of the remaining segregations were significantly different from the expected one (the individual G-tests show a probability of 0.065≤P≤0.997), most of them deviate in the same direction and, hence, the resulting heterogeneity G is clearly not significant (G_H=23.00; P=0.191). The pooled frequencies show this deviation clearly and, being based on a larger sample size, yield a highly significant value of G (G_P=7.00; P=0.008). When all the individual

Table 2. Observed number of Suri and Huacaya offspring in reciprocal Suri × Huacaya crossing.

SURI Sires	Huacaya	Suri	Total
S0270100	3	6	9
S058104	2	9	11
S0810100	1	12	13
S237204	0	7	7
S244203	4	3	7
S443303	7	1	8
SEEI-024	6	9	15
SEEI-025	2	13	15
SSO 502	1	8	9
Total	26	68	94
HUACAYA Sires	**Huacaya**	**Suri**	**Total**
S035104	0	6	6
S095101	3	3	6
S1199-M	4	5	9
S148102	3	6	9
S216204	0	3	3
S322203	3	6	9
S366203	1	7	8
S370397	2	3	5
Total	16	39	55
Grand total	42	107	149

G were added together to obtain a total G, however, the test yielded a value slightly above the critical one (G_T=30.00; P=0.052).

According to the 9:7 segregation hypothesis for complementary gene action (two genes are required for the Huacaya phenotype), 12 progenies out of 19, displayed significantly different results than expected ($0.0001 < P \leq 0.045$) because of an excess of Suri and a low number of Huacaya compared to those expected (Table 5). As all the segregations deviate in the same direction, the resulting heterogeneity G is clearly not significant (G_H=23.22; P=0.182). The pooled G, in view of the consistent trend in favour of Suri, is highly significant (G_P=153.951; P<0.0001); and the total G (G_T=177.388; P<0.0001) is highly significant as well.

On the basis of the total G value, only the latter segregation hypothesis can be completely rejected: the border line values of the total G, resulting from both the two-gene two-alleles (duplicate dominant genes and dominant suppressor genes) models, do not provide any evidence either in favour or against the 15:1 and the 13:3 segregation hypotheses.

Two linked genes

A closer examination of the results of the cross Suri males × Huacaya females in the Quimsachata experimental trial (Table 2), reveals that the segregation ratio seems to be heterogeneous among different males, the extremes being the two males S0810100 and SSO-502 (with one Huacaya in 13 offspring and 1 Huacaya in 8 offspring, respectively), and, on the opposite side, the two males

Table 3. Observed and expected frequencies of Suri × Suri segregating families in a tested 15:1 hypothesis.

Males	Offspring	Observed frequency		Expected frequency[1]		G_{adj}^2	P
		Suri	Huacaya	Suri	Huacaya		
190	32	31	1	29.710	2.290	0.977	0.323
239	18	16	2	16.363	1.637	0.083	0.773
65691	27	26	1	24.954	2.046	0.702	0.402
00-024	27	23	4	24.954	2.046	1.611	0.204
00-034	30	27	3	27.809	2.191	0.291	0.590
0-176	27	26	1	24.954	2.046	0.702	0.402
0-258	19	13	6	17.319	1.681	7.802	0.005
02-848	34	29	5	31.609	2.391	2.376	0.123
1-1-245	6	5	1	4.832	1.168	0.031	0.860
1-2-240	9	7	2	7.723	1.277	0.418	0.518
1-3-214	8	7	1	6.760	1.240	0.058	0.810
1-3-310	12	10	2	10.609	1.391	0.269	0.604
1-3-329	10	8	2	8.686	1.314	0.363	0.547
1-3-333	10	9	1	8.686	1.314	0.093	0.760
2-1-376	19	16	3	17.319	1.681	0.940	0.332
2-3-335	5	4	1	3.867	1.133	0.021	0.885
2-3-342	5	3	2	3.867	1.133	0.749	0.387
9-163	20	19	1	18.276	1.724	0.387	0.534
9-510	44	34	10	41.079	2.921	11.737	0.001
Pooled	362	313	49	339.375	22.625	25.051	0.000

[1]According to the formula suggested by Andresen (1974).
[2]Adjusted G-values according to William's correction.

S443303 and S244203 (with seven Huacaya in 8 offspring and four Huacaya in 7 offspring, respectively). In fact, the G-square statistics of the contingency table were highly significant for heterogeneity (G-square = 28.1, 8 d.f, $P<0.001$). This result justified an analysis in which two hierarchical models were contrasted: (1) the model in which a single parameter R was common to all males; (2) a model in which some males segregated Huacaya offspring at ratio R1 and the other males segregated at ratio R2. The maximum likelihood estimate of R in the first case has already been shown (R=0.290, log likelihood = -20.77). For the second model (only the eight sires with at least one Huacaya offspring were included), the calculated maximum likelihood value separated the first four males, with an estimate of R1 = 0.08, from the other four males, with an estimate of R2 = 0.51, and total log likelihood = -11.76. Since minus twice the log likelihood difference between a model and a more general model is approximately distributed as χ^2, with degrees of freedom determined by the difference in the number of estimated parameters (one in the present case), the first model was rejected in favour of the second ($P<0.001$).

A hypothesis that could explain these unexpected results is that the Huacaya phenotype is determined by the joint homozygosity for recessive alleles at two linked loci, and the Suri phenotype is determined by the presence of a dominant allele at either locus. The double heterozygous animals can be either AB//ab (Cis) or Ab//aB (Trans); if the two loci are tightly linked, only the first diplotype can segregate Huacaya offspring in a test-cross (at a ratio 1:1), and the model is indistinguishable from a single-locus recessive model. If, on the other hand,

Table 4. Observed and expected frequencies of Suri × Suri segregating families in a tested 13:3 hypothesis.

Males	Offspring	Observed frequency		Expected frequency[1]		G_{adj}^2[2]	P
		Suri	Huacaya	Suri	Huacaya		
190	32	31	1	25.992	6.008	7.338	0.007
239	18	16	2	14.543	3.457	0.867	0.352
65691	27	26	1	21.919	5.081	5.628	0.018
00-024	27	23	4	21.919	5.081	0.301	0.583
00-034	30	27	3	24.364	5.636	1.764	0.184
0-176	27	26	1	21.919	5.081	5.628	0.018
0-258	19	13	6	15.367	3.633	1.672	0.196
02-848	34	29	5	27.620	6.380	0.391	0.532
1-1-245	6	5	1	4.421	1.579	0.318	0.573
1-2-240	9	7	2	7.005	1.995	0.000	0.997
1-3-214	8	7	1	6.148	1.852	0.584	0.445
1-3-310	12	10	2	9.547	2.453	0.111	0.740
1-3-329	10	8	2	7.856	2.144	0.012	0.911
1-3-333	10	9	1	7.856	2.144	0.921	0.337
2-1-376	19	16	3	15.367	3.633	0.143	0.705
2-3-335	5	4	1	3.549	1.451	0.213	0.644
2-3-342	5	3	2	3.549	1.451	0.275	0.6000
9-163	20	19	1	16.190	3.810	3.406	0.065
9-510	44	34	10	35.749	8.251	0.434	0.510
Pooled	362	313	49	294.125	67.875	7.003	0.008

[1]According to the formula suggested by Andresen (1974).
[2]Adjusted G-values according to William's correction.

the two loci are separated by a recombination fraction h, both diplotypes can originate Huacaya offspring, the AB//ab diplotype at the ratio R1 = $(1/2 - h)$ Huacaya: $(1/2 + h)$ Suri, and the Ab// aB diplotype at the ratio R2 = $\frac{1}{2}h$ Huacaya: $(1 - \frac{1}{2}h)$ Suri. In fact, if the real Suri males of our crosses are an admixture of the two double heterozygous diplotypes, and if h is not too small, we expect to observe precisely the results of Table 2, with a fraction of the males producing Huacaya offspring at a ratio close to 1:1 and the other fraction producing Huacaya offspring at a ratio close to 0:1.

The above hypothesis prompted us to estimate the recombination fraction h from the data. Assuming that the first four males of Table 2 segregated Huacaya offspring at ratio $\frac{1}{2} h$ and Suri offspring at ratio $1 - \frac{1}{2} h$, and the other four males segregated at ratios $1/2 - h$ and $1/2 + h$, respectively, the maximum likelihood estimate of h was 0.099, with 95% C.L. = 0.029-0.204.

In order to further test this double recessive model of two linked loci using additional independent data, the results of Mallkini records were reanalysed.

Among the 57 males, 34 did not produce any Huacaya offspring, but most of them were not very informative due to their small progeny size. We estimated the segregation ratio of the Huacaya phenotype in this series by maximum likelihood (Table 6). The previous analysis, in which the single-locus recessive model was not rejected, was based on the implicit assumption of a 50:50

Table 5. Observed and expected frequencies of Suri × Suri segregating families in a tested 9:7 hypothesis.

Males	Offspring	Observed frequency		Expected frequency[1]		G^2_{adj}	P
		Suri	Huacaya	Suri	Huacaya		
190	32	31	1	18.000	14.000	28.387	0.000
239	18	16	2	10.125	7.875	9.148	0.002
65691	27	26	1	15.187	11.813	22.986	0.000
00-024	27	23	4	15.187	11.813	10.414	0.001
00-034	30	27	3	16.875	13.125	16.502	0.000
0-176	27	26	1	15.187	11.813	22.986	0.000
0-258	19	13	6	10.687	8.313	1.179	0.278
02-848	34	29	5	19.125	14.875	13.225	0.000
1-1-245	6	5	1	3.289	2.711	2.191	0.139
1-2-240	9	7	2	5.040	3.960	1.864	0.172
1-3-214	8	7	1	4.465	3.535	3.765	0.052
1-3-310	12	10	2	6.745	5.255	4.007	0.045
1-3-329	10	8	2	5.611	4.389	2.528	0.112
1-3-333	10	9	1	5.611	4.389	5.539	0.019
2-1-376	19	16	3	10.687	8.313	6.789	0.009
2-3-335	5	4	1	2.682	2.318	1.514	0.218
2-3-342	5	3	2	2.682	2.318	0.082	0.775
9-163	20	19	1	11.250	8.750	15.555	0.000
9-510	44	34	10	24.750	19.250	8.482	0.004
Pooled	362	313	49	203.625	158.375	153.951	0.000

[1]According to the formula suggested by Andresen (1974).
[2]Adjusted G-values according to William's correction.

Table 6. Segregation of the Huacaya phenotype among half-sib families from Suri × Suri crosses (Mallkini records).

#Huacaya	Number of offspring per male																					Total
	1	2	3	4	5	6	7	8	9	10	12	18	19	20	22	24	27	31	32	34	45	
0	21	1	2		1		1	1		2		1	1		1	1				1		34
1	1				1	1		1		1				1			3		1			10
2				1	2				1	1	1	1										7
3													1					1				2
4																	1					1
6													1							1		2
11																					1	1
Total males	22	1	2	1	4	1	1	2	1	4	1	2	3	1	1	1	4	1	1	2	1	57
Total Huacaya	1	0	0	2	5	1	0	1	2	3	2	2	9	1	0	0	7	3	1	6	11	57
Total offspring	22	2	6	4	20	6	7	16	9	40	12	36	57	20	22	24	108	31	32	68	45	587

proportion of AA:Aa genotypes among the females mated with obligate heterozygous males (selected with at least a Huacaya offspring). Repeating this analysis using an MLE approach (truncated ascertainment model), the following results were obtained: for H = 0.5, R = 0.224 (95% CL 0.172 - 0.330); for H = 0.74, R = 0.165 (95% CL 0.116 - 0.223). Thus, once again, the single-locus recessive model proves to be compatible with the data only for values of H that are incompatible with the observed proportion of segregating males.

Suri offspring on Huacaya × Huacaya segregation

In the Mallkini records, among the 2,126 offspring by Huacaya sire × Huacaya dam, 2,123 were Huacaya and 3 Suri, born from three different families. The few-recorded Suri out of such mating cannot be regarded as recording errors. Also, the hypothesis of a misclassification of heterozygous individuals, showing an incomplete penetrance, cannot be accepted, because the parents of these 3 Suri belonged to pure Huacaya lines. Hence, only the hypothesis of a new dominant mutation on some germinal lines of Huacaya animals can be taken into consideration. The mutation rate of this supposed new dominant mutation can be estimated as 3/2,126 = 0.0014.

Conclusion and future research

In conclusion, the data presented in this work show that the single locus recessive model for the Suri/Huacaya phenotype is unacceptable. In contrast, the data are compatible with the hypothesis that the Huacaya phenotype derives from homozygosity of recessive alleles at two linked loci. This is the simplest model for explaining the results.

Future research will investigate genes involved in the determinism of the phenotypes, based on findings for similar phenotypes in other mammals, the micro-anatomy of the hair follicle and its biology. A candidate gene approach could represent a valid and cost-effective alternative to GWAS and ARFLP genome scanning.

The Fibroblast growth factor 5 (FGF5) gene was shown to be involved through 2 different mechanisms in the determinism of a 'long hair' phenotype in dog, cat, mouse, rat and rabbit (Drogemuller, 2007; Housley and Venta, 2006; Keheler, 2007; Mulsant, 2004; Suzuki, 2000): mutations with loss of functionality and alternative splicing both interfere with the transition between anagen and catagen altering the dermal papilla signalling.

The hair shaft growth rates and the keratinocyte differentiation process are likely to be influenced not only by the signals from the dermal papilla but also by the inner root sheath (IRS) tissue (Van Steensel, 2001). A novel FGF (FGF22) has been isolated in mice by Nakatake *et al.* (2001) and is preferentially expressed in the IRS. Its role as a key molecule in cellular differentiation has been highlighted *in vitro* (Komi-Kuramochi, 2005; Umemori, 2004).

The biological background of the 'lustre' component of Suri fibre is yet to be completely understood. Since keratins are the most important proteins determining physical properties of the fibre, it is likely that their differential expression and that of keratin-associated proteins (KAPs) play an important role, together with the different arrangement of the cuticle cells (Valbonesi, 2010). A recent paper that investigated major genes involved in the determinism of hair phenotypes in dogs has highlighted a mutation on KRT71 (Cadieu, 2009) as responsible for the wave and curly phenotypes in segregating families of Portuguese water dogs. Keratins are obvious candidates from a biological point of view and this specific gene is specifically expressed in the IRS and dominant mutations are known to affect hair structure in mouse and human (Shimomura, 2010; Dunn, 1937; Runkel, 2006).

Beyond keratins and growth factors, there are anatomical factors that could play a major role in the Suri phenotype and hair shape. Asymmetry of the transverse section of the hair shaft/cortex has been shown in other mammals to play a role in crimp and curvature. In particular several authors have demonstrated the bilateral structure (ortho- and para-cortex) of the fibre in sheep with high crimped fibre (merino) as opposed to sheep which do not show this phenotype (Bryson, 2009; Li, 2009; Nagorcka, 1992). In particular some authors have highlighted recently how this asymmetric structure is reflected in mitotical and keratinisation asymmetry thus influencing crimp (Hynd, 2009). Other authors have highlighted the anatomical role of the root sheaths (Thibaut, 2005; Hynd, 2009). Although there are some reports of differences between Suri and Huacaya for hair shaft structure and laterality, the quality of images and results cannot be clearly interpreted.

References

Andresen E., 1974. The effect of ascertainment by truncate selection on segregation ratios. Proceedings 1st World Congress Genetic Applied Livestock Production, Madrid, Spain, pp. 111-114.

Baychelier P., 2000. Suri and Huacaya: Two Alleles or Two Genes? Proc. Australian Alpaca Ass. Nat. Conf., Canberra, Australia, pp 79-85.

Bryson W.G., D.P. Harland, J.P. Caldwell, J.A. Vernon, R.J. Walls, J.L. Woods, S. Nagase, T. Itou and K. Koike, 2009. Cortical cell types and intermediate filament arrangements correlate with fiber curvature in Japanese human hair. Journal of Structural Biology, 166: 46-58.

Cadieu E, M.W. Neff, P. Quignon, K. Walsh, K. Chase, H.G. Parker, B.M.VonHoldt, A. Rhue, A. Boyko, A. Byers, A. Wong, D.S. Mosher, A.G. Elkahloun, T.C. Spady, C. Andre, K.G. Lark, M. Cargill, C.D. Bustamante, R.K. Wayne and E.A. Ostrander, 2009. Coat variation in the domestic dog is governed by variants in three genes. Science, 326: 150-153.

Carmichael I. and G.J. Judson, 1997. Phenotypes resulting from Huacaya by Huacaya, Suri by Huacaya and Suri by Sury Alpaca crossings. Proceedings of the International Alpaca Industry Seminar, Sydney, Australia, pp. 11-13.

Drogemuller C., S. Rufenacht, B. Wichert and T. Leeb, 2007. Mutations within the FGF5 gene are associated with hair length in cats. Animal Genetics, 38: 218-221.

Dunn L.C., 1937. Caracul, a dominant mutation. Journal of Heredity, 28: 334-334.

Housley D.J. and P.J. Venta, 2006. The long and the short of it: evidence that FGF5 is a major determinant of canine 'hair'-itability. Animal Genetics, 37: 309-315.

Hynd P.I., N.M. Edwards, M. Hebart, M. McDowall and S. Clark, 2009. Wool fibre crimp is determined by mitotic asymmetry and position of final keratinisation and not ortho- and para-cortical cell segmentation. Animal, 3: 838-843.

Kehler J.S., V.A. David, A.A. Schaffer, K. Bajema, E. Eizirik, D.K. Ryugo, S.S. Hannah, S.J. O'Brien and M. Menotti-Raymond, 2007. Four independent mutations in the feline fibroblast growth factor 5 gene determine the long-haired phenotype in domestic cats. Journal of Heredity, 98: 555-566.

Komi-Kuramochi A., M. Kawano, Y. Oda, M. Asada, M. Suzuki, J. Oki, T. Imamura, 2005. Expression of fibroblast growth factors and their receptors during full-thickness skin wound healing in young and aged mice. Journal of Endocrinology, 186: 273-289.

Li S.W., H.S. Ouyang, G.E. Rogers and C.S. Bawden, 2009. Characterization of the structural and molecular defects in fibres and follicles of the Merino felting lustre mutant. Experimental Dermatology, 18: 134-142.

Mulsant P., H. Rochambeau and R.G. Thébault, 2004. A note on the linkage between the angora and FGF5 genes in rabbits. World Rabbit Science, 12: 1-6.

Nagorcka B.N. and D.L. Adelson, 1992. The reaction-diffusion system as a determinant of wool fibre diameter, follicle density and fibre diameter distribution. Wool technology and sheep breeding, XL: 47-51.

Nakatake Y., M. Hoshikawa, T. Asaki, Y. Kassai and N. Itoh, 2001. Identification of a novel fibroblast growth factor, FGF-22, preferentially expressed in the inner root sheath of the hair follicle. Biochimica et Biophysica Acta, 1517: 460-463.

Ponzoni R.W., D.J. Hubbard, R.V. Kenyon, C.D. Tuckwell, B.A. McGregor, A. Howse, I. Carmichael and G.J. Judson, 1998. Phenotypes resulting from Huacaya by Huacaya, Suri by Huacaya and Suri by Suri alpaca crossings. Tech. Rep. for Australian Alpaca Assoc. 4 pp.

Presciuttini S., A. Valbonesi, N. Apaza, M. Antonini, T. Huanca and C. Renieri, 2010. Fleece variation in alpaca (*Vicugna pacos*): a two-locus model for the Suri/Huacaya phenotype. BMC Genetics, 11: 70.

Renieri C., A. Valbonesi, V. La Manna, M. Antonini and M. Asparrin, 2009. Inheritance of Suri and Huacaya type of fleece in alpaca. Italian J. Animal Science, 8: 83-91.

Runkel F., M. Klaften, K. Koch, V. Bohnert, H. Bussow, H. Fuchs, T. Franz and M. Hrabe de Angelis, 2006. Morphologic and molecular characterization of two novel Krt71 (Krt2-6g) mutations: Krt71rco12 and Krt71rco13. Mammalian Genome, 17: 1172-1182.

Sambrook J. and D.W. Russell, 2001. Molecular cloning: a laboratory manual, 3rd ed. Cold Spring Harbor Laboratory Press, New York, NY, USA, 999 pp.

Shimomura Y., M. Wajid, L. Petukhova, M. Kurban and A.M. Christiano, 2010. Autosomal-dominant woolly hair resulting from disruption of keratin 74 (KRT74), a potential determinant of human hair texture. American Journal of Human Genetics, 86: 632-638.

Sponenberg P., 2010. Suri and huacaya alpaca breeding results in North America. Small Ruminant Research, 93: 210-212.

Suzuki S., Ota Y., K. Ozawa and T. Imamura, 2000. Dual-mode regulation of hair growth cycle by two Fgf-5 gene products. Journal of Investigative Dermatology, 114: 456-463.

Thibaut S. and B.A. Bernard, 2005. The biology of hair shape. International Journal of Dermatology, 44 Suppl 1: 2-3.

Umemori H., M.W. Linhoff, D.M. Ornitz and J.R. Sanes, 2004. FGF22 and Its Close Relatives Are Presynaptic Organizing Molecules in the Mammalian Brain. Cell, 118: 257-270.

Van Steensel M.A., M. van Geel and P.M. Steiljen, 2001. The molecular basis of hair growth. European Journal of Dermatology, 11: 348-352.

Velasco J., 1980. Mejoramiento Genético de Alpacas. Anales III Reunion Cientifica Animal. Soc. Peruana de Prod. Anim., Lima, Peru.

Producing alpaca fibre for the textile industry

R. Morante[1], A. Burgos[1] and J.P. Gutiérrez[2]
[1]PACOMARCA S.A, Av. Parra 324, Arequipa, Peru; rmorante@pacomarca.com
[2]Dpto. Producción Animal, Facultad de Veterinaria. 28040 Madrid, Spain

Abstract

The textile industry has changed its way of buying fibre, looking for more fine fibre and paying better prices for the finer categories. A change is needed in the type of animal that is bred, e.g. an alpaca that produces more and finer fibre. Traditional methods of genetic improvement have been applied in Peru with no tangible results so it is necessary to use modern methods such as quantitative genetics. The objectives of selection should be focused on the current market, selecting the best animals and reproducing them with advanced techniques achieving its rapid distribution within the alpaca herds of Peru. Pacomarca is an experimental ranch founded by the INCA group with the aim of acting as a selection nucleus from which genetic improvement for alpaca fibre can spread throughout the rural communities in the Peruvian Altiplano. State-of-the-art techniques in animal science, such as performance recording or assisted reproduction including embryo transfer, are applied to demonstrate their usefulness in the Altiplano conditions. Pacomarca has developed useful software (Paco Pro) to carry out the integral processing of production and reproduction data. Mating is carried out individually, gestation is diagnosed via ultrasound, breeding values estimated from a modern genetic evaluation are used for selection and embryo transfer is applied to increase the selection intensity.

Keywords: Alpaca, genetic improvement

Introduction

Raising alpacas in the highlands of Peru is the only source of income for thousands of families. These families make up the lowest social, economic and educational sector. The income for the alpaca breeders is very low and is determined by two key issues that affect the quality and quantity of the product offered to the market: the poor genetics of their animals and the lack of knowledge (education) about modern farming techniques.

The alpaca fibre market has changed substantially over the past ten years. Finer fibre fetches the best market prices to the detriment of courser fibre. There is therefore a great interest among producers to improve the fibre quality produced by their animals. However, most breeders do not know how to produce this kind of fibre, or do not have access to superior genetics that will improve the quality of future offspring.

Due to various social, political and economic factors that have affected Peru in the last 40 years, the size of alpaca herds has fallen (90% are held by small farmers), private investment in breeding has been almost non-existent and the genetics of animals have deteriorated greatly. This genetic problem is the cause of the thickening of the fibre and the very low production rates. However, raising a course fibre alpaca costs the same as raising an improved finer fibre alpaca, while the genetically improved animal can fetch up to 10 times more income. Therefore, the genetic rescue of fine fibre alpacas and greater production efficiency is essential to address this type of breeding in the 21st century, leaving aside the empirical production patterns based on ancient traditions.

The Pacomarca project

In the absence of serious ventures and long-term (state or private) breeding centres for Alpacas in Peru, it was decided to create Pacomarca in the year 2001.

The primary goal of the project was to install a model farm, modern and efficient, which carried out a programme of raising and breeding alpacas using the best technology for producing fine animals and guaranteed the provision of the basics for improving the quality of the national alpaca herd in the hands of small producers. The Pacomarca project objectives can be summarised as follows:
- Genetic improvement focused on the alpaca fibre required by the textile industry.
- Improving animal husbandry techniques allowing better production indices.
- Transmitting the knowledge gained to Andean producers so they can improve alpaca breeding techniques.
- Becoming a genetic core that provides high-quality animal genetics to Andean farmers.
- Research genetics, reproduction, nutrition and other sciences related to the Alpaca.

Genetic improvement programme

The breeding programme of the Pacomarca farm aims to find the quality and quantity of textile fibre needed today, so the selection objectives have been focused on the intended end market where the alpaca fibre goes. To achieve this objective, quantitative genetics techniques were used that have been successful in other species. The stages of the Pacomarca breeding programme are discussed below.

Definition of selection goals

We must select only those traits that are economically relevant for alpacas. In the beginning 26 possible selection objectives grouped were analysed into Fibre Production goals, Biometric goals, Phenotype goals and Yield goals. Taking into consideration the genetic correlations and the importance of each objective, 9 objectives are currently used and divided into two groups (Figure 1): Fibre (objective) and Phenotype (subjective)

The information for objective analysis of fibre is obtained through analysis of fibre sample. Pacomarca animals are analysed once a year by OFDA 100. Until now, 11,553 analyses have been carried out. For the phenotypic information a physical assessment is used at weaning, which

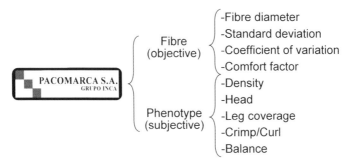

Figure 1. Selection objectives of Pacomarca.

has been standardised for all team members. 3,950 phenotypic assessments have been performed. In Pacomarca 100% of the population for phenotypic analysis and fibre is included.

Performance recording

It is not possible to make genetic improvement if there is no available herd to improve; herds must be systematised and accessible for the project at all times. Pacomarca has developed the PACO PRO software as a main support for the breeding programme (Figure 2). PACO PRO management software can store and organise all data relevant to raising alpacas, saving time and allowing the decision-making process to move on.

The information stored in PACO PRO covers:
- identification and pedigree identification of individual, parents, date of birth, sex;
- reproductive information: breeding, diagnostics, births;
- information on fibre: diameter, standard deviation, comfort factor, coefficient of variation;
- production information: type of fleece, weight, staple length;
- phenotypic information on density, crimp/curl, head, leg coverage, balance;
- medical information: treatments, diseases, defects.

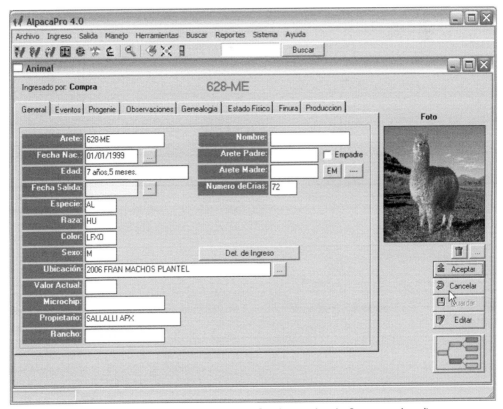

Figure 2. PACO PRO software: individual identification of each animal in the Pacomarca breeding programme.

Estimation of genetic parameters

You can only improve those parameters that have a high heritability. Pacomarca has been able to determine the heritability for the main performance parameters of alpacas (Table 1 and Table 2). Pacomarca has a cooperation agreement with the Univesidad Complutense de Madrid for genetic evaluations of animals based on information from the software Pacopro using restricted maximum likelihood, VCE.

Genetic evaluation

The genetic evaluation (Table 3) is the most reliable statistical estimation of genetic value of a particular individual for a particular trait. Pacomarca uses BLUP (Best Linear Unbiased Predictor) to determine the genetic evaluation of all alpacas in the programme. A genetic evaluation for a

Table 1. Estimated heritabilities (diagonal) and genetic correlations (over the diagonal) for the fibre and phenotype in Huacaya.

	MIC	SD	CONF	CV	DEN	RIZ	CAB	CAL	BAL
MIC	0.3694	0.7194	-0.968	0.0944	-0.079	-0.3	-0.279	0.0279	-0.134
SD		0.417	-0.79	0.7511	-0.257	-0.52	-0.123	0.1105	-0.063
CONF			0.2546	-0.219	0.1076	0.3318	0.2359	-0.077	0.1019
CV				0.3797	-0.296	-0.477	0.1055	0.148	0.059
DEN					0.2364	0.7251	0.2086	-0.145	0.2139
RIZ						0.4198	0.3303	0.0856	0.3672
CAB							0.4254	0.7646	0.9209
CAL								0.4746	0.826
BAL									0.1484

Grey cells: the best alpaca for the trait.

Table 2. Estimated heritabilities (diagonal) and genetic correlations (over the diagonal) for the fibre and phenotype in Suri.

	MIC	SD	CONF	CV	DEN	RIZ	CAB	CAL	BAL
MIC	0.6988845	0.74988	-0.97548	0.08664	0.283993	-0.19308	-0.0349	0.1665	0.0341
SD		0.68429	-0.75836	0.7188	0.141035	-0.15329	-0.01	0.1305	-0.02
CONF			0.56518	-0.1376	-0.33301	0.22377	0.01625	-0.1865	-0.049
CV				0.60515	-0.05584	-0.01802	0.03869	0.0397	-0.052
DEN					0.269231	0.43229	0.71481	0.7919	0.5728
RIZ						0.2854	0.63228	0.4139	0.6877
CAB							0.17598	0.7897	0.9356
CAL								0.3722	0.7287
BAL									0.2585

Grey cells: the best alpaca for the trait.

Table 3. Example of a genetic evaluation chart drawn up for Huacaya males from Pacomarca. Green shows best animal for the trait and red the worst. Male 178-05M is son of male 17M showing the genetic gain.

ARETE	DOB	MIC VG	SD VG	FCO VG	CV VG	DEN VG	RIZ VG	CAB VG	CAL VG	BAL VG
505-07M	07-03-07	**-2.39572**	-0.13734	4.55725	-0.10795	0.27155	0.49527	0.29288	0.28385	0.33169
549-06M	10-05-06	-2.22733	-0.25671	3.52586	0.69244	0.26001	0.26227	0.50187	0.29505	0.26631
373-05M	01-03-05	-2.12048	-0.16223	4.24445	0.56557	0.24192	0.30132	0.22031	0.16533	0.18304
557-06M	14-05-06	-2.02425	-0.30582	1.57401	0.10927	0.01458	0.63657	0.10591	0.20374	0.02644
35M	07-03-00	-1.83636	-0.54211	4.64751	-1.30565	0.21939	0.53465	0.24074	0.24715	0.22842
44M	19-01-04	-1.73619	-0.52152	1.87182	-1.21075	0.02911	0.18504	-0.26624	**-0.24144**	-0.08362
436-07M	17-02-07	-1.53701	-0.55877	1.77469	-1.24383	0.16331	0.30021	0.85083	0.38175	0.28574
309-07M	23-01-07	-1.49774	-0.23280	1.43647	-0.29754	0.12211	0.15933	0.39515	0.51422	0.22272
374-05M	01-03-05	-1.47652	-0.57071	0.74719	-0.70127	0.69799	*1.31615*	0.39915	0.47308	0.39930
122-07M	15-12-06	-1.47576	**-0.64508**	2.51595	-1.01303	0.16213	0.20088	0.12658	-0.19402	0.15112
079-06M	29-12-05	-1.42677	0.15371	4.75618	0.62861	0.05061	0.41967	0.13262	0.25080	0.23967
34M	07-01-00	-1.27016	-0.37026	1.54858	-0.52700	0.17085	0.20216	0.19840	-0.05787	0.12095
374-06M	16-02-06	-1.25376	-0.18887	-1.17160	-0.30822	0.44500	0.39631	0.69042	0.35212	0.33450
124-06M	07-01-06	-1.19700	-0.34699	1.78120	-0.81013	-0.05477	0.37608	-0.13841	0.18406	-0.05188
619-07M	23-04-07	-1.17759	-0.18832	-0.11902	0.46401	0.05593	0.26341	0.18207	0.32550	0.13857
043-02M	16-01-02	-1.06246	0.13772	-0.79671	0.66705	0.39888	0.31654	0.54022	0.06511	0.35420
178-05M	14-01-05	-1.03305	0.21704	0.77700	1.11221	-0.14584	-0.03153	0.15731	0.59750	0.15044
303-06M	06-02-06	-0.98028	-0.41995	0.87505	-0.38156	0.40957	0.42464	-0.06384	-0.21983	-0.08168
293-06M	04-02-06	-0.85667	-0.40015	1.65770	-0.27741	0.48571	0.56126	0.34411	0.04476	0.11338
6M	15-02-95	-0.83755	-0.48089	4.02589	**-2.20956**	0.13748	0.25957	0.12792	-0.05056	-0.03551
213-05M	21-01-05	-0.68192	-0.15531	0.75239	0.24303	0.43266	0.64204	0.11209	0.48520	0.21231
553-06MR	11-05-06	-0.59486	0.00145	-0.10427	1.06636	0.02501	**-0.19898**	**-0.45627**	0.03498	**-0.16376**
12M	01-01-00	-0.44560	-0.44342	0.63863	-0.17334	0.08328	0.21790	-0.22352	-0.10493	0.01301
004-01M	27-12-00	-0.44067	-0.25648	-1.34579	0.16313	0.61239	1.08804	-0.17538	-0.00931	0.13106
363-06M	14-02-06	-0.43264	-0.48333	1.68519	-0.55005	0.19950	0.16608	-0.41273	-0.04051	-0.05885
053-02M	24-01-02	-0.35441	0.07461	**-1.89626**	0.77514	-0.10833	-0.14478	0.74816	0.98439	0.50068
17M	23-01-00	-0.32551	0.61089	-1.51371	3.09721	-0.46094	-0.05667	0.37561	0.84230	0.30379
207-06M	20-01-06	-0.26960	0.06364	-0.41885	0.07811	0.53355	0.81455	0.33184	-0.05798	0.03971
059-04M	25-12-03	-0.21110	-0.32211	-0.34557	0.06727	0.64676	0.58272	-0.06305	-0.04006	0.12865
454-06M	15-03-06	-0.20673	-0.21479	-0.72318	-0.38947	0.16373	0.28865	0.49716	0.49083	0.28946
026-05M	16-12-04	-0.02907	-0.58663	0.90629	-1.41583	0.11496	-0.15623	0.64662	*1.00021*	0.40586

Bold: the best alpaca for the trait.

Italic: the worst alpaca for the trait.

Grey cells: male 178-05M is son of male 17M showing the genetic gain.

given trait is the expected difference between the average value of the offspring of an individual and the average value of the trait in the rest of the population. The value lets you determine how much better or worse an animal is with respect to the mean for each trait, allowing the selected alpacas the transmission of the best production traits to their offspring.

Genetic testing is carried out every year for all animals, based on these results; breeding decisions are made, where the male complement the lack of females. Likewise with the results of genetic evaluation animals can be selected for entering the embryo transfer programme.

Selection criteria

The two main tools of genetic improvement are the selection (to determine which individuals are going to leave offspring) and mating systems (to determine how the individuals selected will be paired). Breeding involves the processes of genetic evaluation and dissemination of selected genetic material in which you can use artificial reproductive technologies such as embryo transfer (ET).

Pacomarca uses the following breeding techniques:
- Individual breeding: for which replicated facilities designed to facilitate this work achieved 92% fertility.
- Embryo transfer: after the experience of genetic evaluations, Pacomarca has acquired the ability to identify their best animals. However, a rapid genetic response is limited by the number of offspring produced naturally, only one offspring born to each female per year. To address this limitation, Pacomarca has initiated a programme of assisted reproduction through embryo transfer. Males and females with the best breeding values were selected. Up to six embryos, with an average of four, were obtained from each female and transferred to females with high maternal capabilities. Thus, each female can provide four offspring per year on average. Table 4 shows the summary of the data for embryo transfer in the 2009-2010 season, showing 68% of successful pregnancies and 60% of offspring born.

Table 4. Summary of embryo transfers production in 2009-2010.

	n	%
Donor females	25	-
Recipient females	145	-
Flushings	131	-
Collected embryos	101	77.10
Transferred embryos	101	77.10
Successful pregnancy	69	68.32
Offspring born	61	60.40

Meat and fibre production

External effects for the quantity and quality of fibre in Guanacos

J. Von Thüngen[1], R. Lheure[2], E. Seguineau De Preval[2] and A. Perazzo[1]
[1]*Instituto Nacional de Tecnología Agropecuaria Estación Experimental Agropecuaria Bariloche, Modesta Victoria 4450, 8400, S.C. Bariloche, Río Negro, Argentina; jvthungen@bariloche.inta.gov.ar*
[2]*AgroParisTech 16 Rue Claude Bernard F-75005, Paris, France*

Abstract

The study is part of a project developing sustainable use for the guanaco (*Lama guanicoe*). The guanaco is a wild camelid living free on the Argentine Patagonian farms. Once a year, the animal is live captured, shorn and released, and its fibre sold. The objective is to define how the amount and the quality of fibre obtained per animal can vary. The impact of some external factors on these variables of economic interest has to be determined, which is the content of this study. Sampled guanacos were trapped live, shorn and permanently tagged between 2005 and 2008, fleece weighed and fleece samples of 10 cm^2 were taken from the shoulder during the captures. The analysed variables were mean fleece weight (MFW), average fibre diameter (AFD), and down yield (DY). AFD and DY were obtained using Sirolan-Laserscan (IWTO12) and the Hermann and Wortmann (1997) method validated by Sacchero and Mueller (2005) respectively. All the data have been compared statistically with mean-tests and mean-tests in pairs ($P<0.05$). Season, rainfall and previous shearing were analysed. Significant differences were detected for rainfall and previous shearing. No significant differences were found for season. Little information was available about the quality: only the effect of rainfall could be analysed. The results tend to show a constant yield and a smaller diameter of fibre in dry conditions as happens with other species. However, it is necessary to continue monitoring these effects as guanaco could be a good use of dry territories in Patagonia.

Keywords: guanaco, quality, quantity, climate

Introduction

Argentina is covered by 70% of arid and semi-arid land. Production is limited by paucity and variability of rainfall. In Patagonia, the main production system has been sheep farming since the beginning of the twentieth century. The system is built on the production of wool, which is essential to the textile industry (Von Thüngen and Lanari, 2010; Mueller, 2005). In 2008, significant quantities of animal fibres were registered in Argentina: 54,000,000 kg of wool, 70,000 kg of Llama, 825,000 kg of Mohair, 2,150 kg of Cashmere, 377 kg of Vicuña and finally 1,500 kg of Guanaco (INTA National Program of Animal Fibres, 2009).

The exploitation of the land today leads to several problems. Traditional sheep breeding has often been managed without any consideration for the environment (overgrazing by domestic herbivores, habitat alteration by deforestation and uncontrolled fires) (Baldi *et al.*, 2001; Defossé *et al.*, 1992). Consequently land degradation, larger problems of erosion and endangered species are the legacy in Patagonia. However, livestock keepers, land users and the community at large are exploring sustainable opportunities for the use of these lands (Baldi *et al.*, 2006).

The guanaco (*Lama guanicoe*) is the most conspicuous wild ungulate of Patagonia. However, the actual running hypothesis is that populations in South America, including Argentina, have declined continuously since European settlement up today (Torres, 1985; Cunazza *et al.*, 1995). Causes for this decline are related to habitat alteration, competition for forage by sheep (Baldi *et al.*, 2004), hunting – formal and informal – and lack of reasonable management plans to develop simultaneously a livelihood for ranchers and for the needs of conservation (Baldi *et al.*, 2006; Von Thüngen and Lanari, 2010). Today, the idea is to apply 'the use it or lose it' concept (Franklin and Fritz, 1991), by using the fibre. The guanaco has a very dense hair that protects it. This fibre provides appreciated characteristics in the textile industry. As the guanaco is a wild animal, it is necessary to develop corral traps to capture it alive before shearing and releasing it.

In order to optimise the use of the guanaco the external factors affecting the quality and quantity of fibre have to be known. The quality can be evaluated by several parameters: yield, average diameter of fibre, staple length, resistance to traction. The higher the yield and thinner the fibre, the higher the price. The quantity of fleece, expressed in grams, is the weight of fibre obtained after shearing. Three factors have been chosen to be studied: the effect of the season (the best time of year to shear the guanaco), the effect of a previous shearing and finally the effect of the climate.

This article studies the sustainable use of guanaco in Patagonia. In order to define the economic aspect, the influence of external variables on parameters of interest – quantity and quality of fleece collected – has been analysed.

Materials and methods

Data used for the analysis were collected from a wild guanaco population living on a 25,000 ha farm near Piedra del Águila (Latitude 40 °S and Longitude 70 °W) in the province of Neuquen. The climate is semi-arid with dry a season in summer, and rainfall occurring throughout cold winters. Annual precipitation varies between 250-400 mm and the average annual temperature is 7 °C (Bustos and Rocchi, 1993). Landscape is shaped by vast plains of shrub land and tussock grasses ranging around 900 m altitude (Ayesa *et al.*, 1995; Defosse *et al.*, 1992; Elissalde *et al.*, 1995).

The wild guanaco population is trapped live, shorn and released every year. About 1000 animals a year are captured in various capture sessions. As guanacos are territorial, some animals are recaptured in successive years. There are two possible moments to capture guanacos: October-November and February-March. Some years, the captures are only done during one season; the fleece samples have been collected from 2005 to 2007, during the summer sessions. These guanacos were permanently tagged between 2005 and 2008. For some of them, the fleece was weighed and sampled during the captures. Fleece samples of 10 cm^2 were also taken from the shoulder according to Defosse *et al.* (1980). Average fibre diameter (AFD) was obtained using Sirolan-Laserscan (IWTO12). The yield of down fibres was estimated by the Hermann and Wortmann (1997) method validated by Sacchero and Mueller (2005). Table 1 lists the available data.

The probability of capturing one guanaco is mainly related to the experience of the personnel involved, but weather conditions also influence the amount of animals captured at any one time. Probability of recapturing one guanaco is related to its experience. Every guanaco is identified by a mark (tag). We classed the animals in groups, according to their history of capture as shown in Table 2.

To analyse the effect of the season, comparison test by pairs was done for animals caught two or three times between 2005 and 2006 (Daudin *et al.*, 2001). The study of the climate effect was

Table 1. Number of fleece sampled from successive capture events, resulting in availability of raw data.

Year	Summer-session	Spring session	Fleece sample recollection
2005	384	-	-
2006	1,093	735	30
2007	617	-	83
2008	153	-	102

Table 2. Number of guanacos captured with same capture history.

Summer 2005 Feb-Mar	Summer 2006 Feb-Mar	Spring 2006 Oct	Total number
1	0	0	240
1	1	0	55
1	0	1	72
1	1	1	17
0	1	0	950
0	1	1	71
0	0	1	575

done by comparing with a mean test the weight of fleece for all the animals captured in 2005, 2006 and 2007 (Daudin *et al.*, 2001). A correlation test was run with rainfall data available in the region (Bustos and Rocchi, 2008).

To understand the effect of a previous shearing we compared AFD and DY for animals shorn for the first time and second time. To avoid a bias due to the year effect we analysed animals shorn in the same year. (Summer 2006 with group Summer 2005-Summer 2006). All the statistics have been considered with a fixed α error of 5%.

Results and discussion

Effects on the quantity of fibre

Season effect

The mean-tests were done on the following group associations: Su5-Su6; Su5-Sp6; Su6-Sp6 without G1; Su6-Sp6. They show that the only significant difference is between the late summer 2005 and the beginning of spring 2006 ($\mu_{Su5}=\mu_{Su6}$; $\mu_{Su6}>\mu_{Sp6}$; $\mu_{Su6}>\mu_{Sp6-G1}$; $\mu_{Su5}=\mu_{Sp6}$). The weight of fleece seems to be equal, whatever the season, except for the guanacos shorn for the second time in the same year (2006). This can be explained by shorter lapses of time between the two captures in question (eight months instead of one year).

Year effect

The year effect was analysed for four consecutive years; between which mean-tests have been carried out as shown in Table 3. There is a significant difference in fleece weight between 2005 and for all the other years. For all other comparisons, no significant difference was observed.

To explain this point we looked at the rain data for these years. There was a shortage of rainfall in 2003 and 2004. This could have provoked poor body condition at the end of 2004, causing a reduction in the growth of the fibre for this year. Finally it could explain the small weight of fleece in January 2005 in comparison with the other years. However, these results seem to be in contradiction with the previous tests where no significant difference was found.

Shearing effect

To complete this study, we also tested shearing effect, to evaluate differences in fleece weight between first and second shearing (Table 4). To avoid a bias due to the year effect, we compared a set of animals shorn for the first time in 2006 and another set of animals shorn for the second time in the same year.

Year effect through mean-test shows only significant differences between 2005 and the other years. However, there is a significant shearing effect between first and second shearing in fleece weight. The first shearing produces more fibre than the second one.

Effects on the quality of the fibre

As this is a wild population and the history of capturing is recent, only the climate effect on the quality of the fibre could be studied (Figure 1).

The AFD seems to follow the same evolution as fleece weight. Mean-tests on the AFD are shown in Table 5. The DY is nearly constant over the years, except in 2007, where it is smaller.

Table 3. Year effect on the quantity of fibre.

	Group A6	Group A7	Group A8
Group A5	$\mu5<\mu6$	$\mu5<\mu7$	$\mu5<\mu8$
Group A6	-	$\mu6=\mu7$	$\mu6=\mu8$
Group A7		-	$\mu7=\mu8$
Group A8			-

Table 4. Average fleece weights.

	Number of guanacos	Average	Variance	97.5 percentile	Statistic
Second shearing	55	279.44	7,697.47	1.96	2.46
First shearing	1,007	309.54	9,459.86		

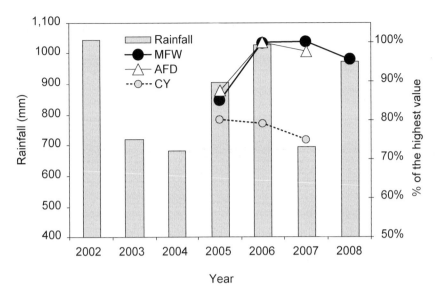

Figure 1. Variation of key parameters with rainfall.
MFW: mean fleece weight; AFD: average fibre diameter; DY: down yield.

Table 5. Differences in average fleece weight.

AFD	2006	2007
2005	$\mu_{2005}<\mu_{2006}$	$\mu_{2005}<\mu_{2007}$
2006		$\mu_{2006}=\mu_{2007}$

Conclusions

The guanaco is an animal that provides a good quality fibre and the use of this fibre might offer an opportunity for its conservation. This first study shows that the time at which the fleece is collected does not seem to affect the quantity or the quality of the fibre. It can be concluded that there is no seasonal effect. However, further detailed studies in the evolution of moulting may indicate the best moment to shear. The climatic conditions seem to have an impact on fleece weight and AFD in a dry context. The resistance at traction in relation to climatic changes should be analysed in the future to confirm these initial conclusions.

The ability to predict variations in quality fibre according to climatic and range conditions would enable farmers to decide whether and at what moment to invest in capturing guanacos. If the animals are in a bad condition, for example after a shortage of water, they produce less fibre, but of a better quality. This is very interesting because it leads to the conclusion that the guanaco is an animal able to make optimal use of poor land, on which other cattle could not survive. Studies for other species have already been conducted.

Acknowledgements

The authors wish to thank Estancia La Esperanza SRL and all its personnel for field assistance and collaboration in obtaining samples. The study was supported by two INTA projects: AERN 292211 and PNFIB 131931.

References

Ayesa J., D. Bran, C. Lopez, S. Cingolani, S. Clayton and D. Sbriller, 1995. Evaluación del estado actual de la desertificación en la transecta Río Negro, Evaluación del Estado Actual de la Desertificación en Áreas Representativas de la Patagonia, Informe Final de la Fase I. Proyecto de Cooperación Técnica entre la Republica Argentina y la Republica Federal de Alemania, Rio Gallegos, Trelew, Puerto Madryn, Bariloche, Argentina.

Baldi R, S.D. Albon and D.A Elston, 2001. Guanacos and sheep: evidence of continuing competition in arid Patagonia. Oecologia, 129: 561-570.

Baldi R., D. del Lamo, M. Failla, P. Ferrando, M. Funes, P. Nugent S., Puig, S. Rivera and J. Von Thüngen, 2006. Plan Nacional de Manejo del guanaco (*Lama guanicoe*). (National Management Plan for guanaco (*Lama guanicoe*) Secretaria de Ambiente y Desarrollo Sustentable de la Nacion. Argentina.

Bustos J.C. and V.C. Rocchi, 2008. Caracterización termopuvliometrica de veinte estaciones meteorológicas de Rio Negro y Neuquen. Temperature recordings from 20 meteorological stations of Rio Negro and Neuquen. INTA-EEA Bariloche. Area de Recursos Naturales. Agrometeorología. Pub. Técnica, 26: 43.

Cardellino R.C. and J.P. Mueller, 2008. Fibre production and sheep breading in South America. Proceedings Association for the Advancement of Animal Breeding and Genetics 18: 366-373.

Cunazza C. and S. Puig, 1995a. Situación del guanaco y su ambiente. (Situation of the guanaco and its environment) In: S. Puig (ed.), Técnicas Para el Manejo del Guanaco, Gland. IUCN/SSC South American Camelid Specialist Group, Switzerland, pp: 27-50.

Cunazza C., F. Videla and N. Oporto, 1995b. Manejo para recuperación de poblaciones en peligro. In: Puig, S. (ed.), (Management for recovery of endangered species), Técnicas Para el Manejo del Guanaco, Gland. IUCN/SSC South American Camelid Specialist Group, Switzerland, pp: 171-177.

Daudin J.J., S. Robin and C. Vuillet, 2001. Statistique inférentielle; Idées, démarches, exemples (inferential statistics; ideas, approaches, examples). Presses Universitaires de Rennes, Rennes, France.

Defosse A.M, J.L. Garrido, O.J. Laporte and L. Duga, 1980. Cria de guanacos en cautividad: variación de su crecimiento y calidad de su lana. (Breeding of guanacos in captivity: variability of their growth and quality of their fibre) Estudio de fauna silvestre de interes económico. Programa de Ecología y Desarrollo Regional de Zonas Aridas y Semiáridas. Centro Nacional Patagonico, Pueto Mandryn, Argentina, 17 p.

Defossé G.E., M.B. Bertiller and C. Rostagno, 1992. Rangeland management in Patagonian drylands. In: G.K. Perrier and C.W. Gay (eds.), Proceedings of the International Rangeland Development Symposium, Society for Range Management, Spokane, WA, USA, pp. 12-21.

Elissalde N.O., J.E. Pappalardo and M.S. Mühleman, 1995. Evaluación del estado actual de la desertificación en la transecta Chubut, Evaluación del estado actual de la desertificación en áreas representativas de la Patagonia, Informe Final de la Fase I. Proyecto de Cooperación Técnica entre la Republica Argentina y la Republica Federal de Alemania. Rio Gallegos, Trelew, Puero Madryn, Bariloche, Argentina.

Frank E.N., M.V.H. Hick, H.E. Lamas, C.D. Gauna and M.G. Molina, 2006. Effects of age-class, shearing interval, fleece and color types on fiber qualities and production in Argentine Llamas. Small Ruminant Research, 61: 141-152.

Franklin W.L. and M.A. Fritz, 1991. Sustained harvesting of the Patagonia guanaco: Is it possible or too late?. In: J.G. Robinson and K.H. Redford (eds), Neotropical Wildlife Use and Conservation, University of Chicago Press, Chicago, IL, USA, pp. 317-336.

Mueller J.P., 2005. Introducción a la producción ovina Argentina. In: J.P. Mueller and M.I. Cueto (eds.), Actualización en Producción Ovina 2005. Memorias VII Curso de Actualización en Producción Ovina. San Carlos de Bariloche, Río Negro, Argentina.

Rabinovich J.E., M.J. Hernández and J.L. Cajal, 1985. A simulation model for the management of Vicuña Populations. Ecol. Modelling, 30: 275-295.

Rabinovich J.E., J.L. Cajal, M.J. Hernández, S. Puig and J. Amaya, 1984. Publicación de la División de Recursos Naturales Renovables, Subsecretaría de Ciencia y Tecnología, Buenos Aires, Argentina.

Sacchero D.M. and J.P. Mueller, 2005. Determinación de qualidad de vellones de doble cobertura tomando al vellón de Vicuña (*Vicuña vicuña*) como ejemplo (Determination of the quality of fleece with double cover, taking the fleece of Vicuna as an example). RIA 34: 143-159.

VonThüngen J. and M.R. Lanari, 2010. Profitability of sheep farming and wildlife management in Patagonia. Pastoralism, 1: 274- 290.

Torres H., 1985. Distribución y conservación del guanaco (Distribution and conservation of the guanaco). Special report No. 2 IUCN/SCC.

Wurzinger M., J. Delgado, M. Nürnberg, A.V. Zarate, A. Stemmer, G. Ugarte and J. Sölkner, 2006. Genetic and non-genetic factors influencing fibre quality of Bolivian llamas, Small Ruminant Research 61: 131-139.

Genetic and non-genetic factors influencing fibre quality of Arkharmerino × Ghezel and Arkharmerino × Moghani crossbreeds of sheep in third generation

H. Esfandyari[1], A.A. Aslaminejad[1] and S.A. Rafat[2]
[1]*Department of Animal Science, Faculty of Agriculture, Ferdowsi University of Mashhad, Mashhad, P.O. Box 9177948974, Azadi Square, Mashhad, Iran; Esfandyari.Hadi@gmail.com*
[2]*Department of Animal Science, Faculty of Agriculture, University of Tabriz, Tabriz, Iran*

Abstract

This investigation was carried out to reveal the comparative production performances and effect of some environmental factors on the wool characteristics of Arkharmerino × Ghezel and Arkharmerino × Moghani crossbred sheep in third generation. The research material consisted of fleece samples taken during a period of two years (2007-2008) from the midside region of the animals. Each sample was measured for average fibre diameter, fibre diameter variability, staple length, proportion of medullated fibre, proportion of kemp and comfort factor. The comparative values for these fleece characteristics in F3th Arkharmerino × Ghezel were 28.78±0.48μ, 36.84±1.16%, 11.94±0.35cm, 7.07±0.93%, 1.02±0.23% and 68.93% and for, Arkharmerino × Moghani were, 29.79±0.43μ, 41.86±1.16%, 11.96±0.37 cm, 8.13±1.06%, 2.71±0.45% and 63.33±3.66% respectively. The effects of genotype, sex, birth type, and year of birth were studied. Genotype had a significant ($P<0.01$) effect on average fibre diameter and proportion of kemp and Arkharmerino × Ghezel crossbreeds had lower diameter with a smaller proportion of kemp. The two differences in fibre characteristics that were attributable to sex were fibre diameter variability and proportion of medullated fibre and females had higher amounts than males for both trait. As a result of the statistical analysis, it was found that crossing with Arkharmerino generally had positive effects on the fleece, favouring its use in the hand-woven authentic carpet industry.

Keywords: carpet wool, Arharmerino, Moghani, Ghezel, crossbreed

Introduction

Sheep breeding plays an important role in animal production in Iran. Sheep have the ability to transform poor grasslands, which are wide-spread in Iran, into valuable products like meat, milk, wool and skin. In Iran, sheep meat, sheep milk and products are valuable and generally preferred commodities. The value of indigenous breeds is in their ability to continue production under poor feeding and management conditions and in the possibility that their genetic make-up may fit new environmental conditions which might occur in the future (Yilmaz *et al.*, 2003). The indigenous Ghezel and Moghani sheep breeds are raised in the north-east of Iran. In this region these two sheep breeds have been around for more than a hundred years. The relatively more favourable topography and denser population have contributed to make the region one of the most intensively cultivated in the country. Therefore, sheep in this region have to compete with field crops.

Most of the wool produced by the indigenous sheep breeds in Iran is of coarse and mixed type wool used in the hand-woven authentic carpet industry. It is reported that there are 240 thousand hand-woven carpet production workbenches spread throughout Iran and about 5.1 million m² hand-woven carpet is being produced annually. This is an important occupation, particularly for the women living in rural areas. The authentic hand woven carpet type of wool is preferred if it is coarse and mixed type as the thick fibre helps the thread to stand upright in the carpet (Demir and

Baspinar, 2001). Even though native breeds of Ghezel and Moghani produce wool that is useful for the carpet section, uniformity of the wool diameter along the staple and different regions of body is high and should be reduced. In addition, these two breeds have the potential to produce finer wools to meet the Fustian system qualifications (Shodja *et al.*, 2004). Crossbreeding with the exotic Arkharmerino breed to reach these goals started in the region during 2001.

Many crossbreeding experiments have been conducted to estimate the genetic effects of breeds of sheep and, typically but not always, increased lamb production has been the primary objective of such experiments (Wolf *et al.*, 1980; Kempester *et al.*, 1987). Nevertheless, comprehensive evaluation requires that other factors, such as wool production and quality, also be considered. Wool is an additional source of income in most sheep operations and can represent 15 to 25% of gross income (Cedillo *et al.*, 1977). Therefore, breed effects for traits that influence fleece value should be considered when selecting breeds for a crossbreeding system. In this object previous papers in this series have described wool characteristics of Ghezel and Moghani ewes when mated with Arkharmerino rams and their crosses (Mokhber, 2005). The objective of the current study was to evaluate wool characteristics of third generation Arkharmerino × Ghezel (ArGh) and Arkharmerino × Moghani (ArMo) and describe the genetic and environmental effects on fibre quality of crossbreeds.

Materials and methods

The present study is part of a general project to improve sheep production by crossbreeding with fine wool breeds in the north-west of Iran, a region which borders Turkey in the west and with Azerbaijan in the north. The area has moderate summers with humid and cool winters and a moderate amount of snow. The average annual temperature is 6.5 to 15 °C. The third generation of ArGh and ArMo crossbreed sheep were produced and maintained at the Khalat Poshan Research Station affiliated to the University of Tabriz located in Tabriz, Iran. The wool samples of the ArGh and ArMo genotypes, which were collected in the station, were used in the study. The two genotypes were managed in the same flock in the station. Both male and female flock were kept indoors during the winter months and then taken to pasture as the weather conditions improved. In the station shearing was done once a year, usually at the end of June. Before shearing, wool samples were taken from the side (from the rib region) of 154 ArGh and 142 ArMo crossbred sheep as presented by sex and genotype in Table 1.

The following objective measurements were conducted on each sample using the referenced international or national standard methods in Animal Fibre Technology Ltd, Tabriz University, Iran. Each raw sample was measured staple length (SL) by ruler on velvet and then blended, scoured, dried and conditioned in a standard atmosphere (20±2 °C and 60±2% RH) for other measurements. The entire sample was used. Snippets of 2 mm length were cut with a Heavy Duty Sectioning Device (guillotine) and spread on an open glass slide and then the slides were analysed by computer-assisted Image Analysis method. The physical components of the instrument for taking measurements consisted of a web maker, an image acquisition system and a computer for image processing. Proprietary software was used to analyse fibre images stored

Table 1. Number of F_{3s} crossbreeds by genotype and sex.

ArGh	Male	70
	Female	84
ArMo	Male	71
	Female	71

in the computer memory. The following traits were analysed: average fibre diameter (AFD), fibre diameter variability (FDV), proportion of medullated fibres (PMF %), proportion of kemp fibres (PK %) and comfort factor (CF %). After determination of the average fibre diameter, proportion of medullated, kemp and comfort factor were measured in the same sample produced for each sheep. The proportion of medullated fibre (% by number) is defined by Projection Microscope as the proportion of fibres with less than 60% cannel in fibre. Data for medullated fibres were only analysed in white and less-coloured samples, because the Projection Microscope seems to overestimate the proportion in coloured fibres (Delgado *et al.*, 1999). Medullated fibres are hollow and large in diameter and are stiff and harsh. They also appear lighter coloured when dyed. Therefore, medullated fibres are undesirable. The proportion of kemps (% by number) is defined by the projection microscope as the proportion of fibres with more than 60% medulla in fibre. Kemp is a technical term referring to a fraction of medullated fibres. The separate mention of kemps was established by the American Society for Testing and Material Standards (ASTAM's Committee D-13), which recognised that kemps are the source of more visible problems in fibre processing than medullated fibre (McGregor and Butler, 2004) and the percentage for kemp, medulla and comfort factor is determined with a count of 1000 fibres in each sample. FDV was measured by calculating the coefficient of variability for each wool sample (Fahmy, 1987).

Traits were tested for normal distribution and transformed when necessary. Only the comfort factor showed a normal distribution. A logarithmic transformation ($y = \log_{10} x$) was performed for AFD and SL. PK and PMF was transformed with the function of Box and Cox ($y = ((x^{0.2} -1)/0.2) -1$) (Delgado, 2003). The following statistical model was used for estimation of fixed effects:

$$Y_{ijkl} = \mu + GP_i + S_j + Tb_k + Yb_l + e_{ijkl} \tag{1}$$

where Y_{ijkl} refers to an observation on an individual animal; μ is the constant common to all individuals; GP_i the fixed effect of genotype, i=1, 2 (ArGh and ArMo); S_j the fixed effect of sex, j=1, 2 (male, female); Tb_k the fixed effect of birth type, k = 1, 2 (single and twin); Yb_l the fixed effect of year of birth, l = 1, 2 (2007 and 2008), and e_{ijkl} is the residual effect.

The general linear model procedure (PROC GLM) of SAS (SAS Institute Inc., Cary, NC) was used to calculate and separate means (Duncan's multiple range test) and least square means for the 6 characteristics recorded.

Results and discussion

It is difficult to compare the results from the literature with the results presented here. Different methods such as air flow, Projection Microscope or OFDA are used in laboratories for determining fibre traits. It is not always clear whether published figures come from complete or de-haired fleeces. In addition, definitions for some traits such as kemp or medullated fibre vary. Extreme care must be taken to select comparable data sets (Lupton *et al.*, 2006). Mean, standard deviation, minima and maxima for measured traits (AFD, SL, FDV, PK, PMF and CF) have been presented by genotype (ArGh and ArMo) in Table 2.

The average fibre diameter in ArGh and ArMo in the third generation was 28.78 µm and 27.79 µm respectively. The fibre diameter of two genotypes is intermediate and comparable to carpet-wool breeds, 31 µm for Barki sheep, 26.2 µm for Arabi sheep, 35.4 µm for Ossimi and 31.5 µm for Rahmani (Tabbaa *et al.*, 2001). Shodja *et al.* (2004) reported 34.7 and 27.53 µm for Ghezel and Moghani pure breeds respectively. FDV in ArGh and ArMo were 36.84% and 41.86% respectively. The FDV in carpet wool breeds varies considerably 27-55% and the present data

Table 2. Mean, standard deviation (SD), minimum and maximum for measurements by genotype.

Genotype[1]	Measurement[2]	No.	Mean	SD	Min	Max
ArGh	AFD(μm)	154	28.78	3.55	22.23	36.99
	FDV (%)	154	36.84	8.54	22.06	57.40
	SL (cm)	154	11.96	2.70	5	18.50
	PK (%)	154	1.02	1.73	0	8.1
	PMF (%)	154	7.07	6.91	0	27.6
	CF (%)	154	68.46	13.82	34.25	92.75
ArMo	AFD(μm)	142	27.79	2.83	23.62	35.71
	FDV (%)	142	41.86	10.35	18.72	67.66
	SL (cm)	142	11.56	2.24	7	16.10
	PK (%)	142	2.71	2.96	0	14.4
	PMF (%)	142	8.13	5.80	0.62	31.4
	CF (%)	142	63.26	11.7	35.71	87.64

[1]ArGh: Arkharmerino × Ghezel; ArMo: Arkharmerino × Moghani
[2]AFD, average fibre diameter; FDV, fibre diameter variability; SL, staple length; PK, proportion of kemp; PMF, proportion of medullated fibre; CF, comfort factor.

falls within this range. Shodja *et al.* (2004) results for this trait in pure breeds were 39.84% and 44.19%, respectively. The wide-ranging FDV in this study reflects the absence of selection for this trait in past generations. Lupton *et al.* (2004) reported FDV in Finewool sheep breed of about 7 μm. It shows that FDV in studied crossbreeds are very high and it must be reduced by selection.

Table 2 shows that staple length in ArGh and ArMo was 11.96 cm and 11.56 cm respectively. The observed data in the present study is smaller than the 13-20 cm range in other Middle Eastern countries. Shodja *et al.* (2004) reported staple length for Ghezel and Moghani breeds 12.37cm and 11.72 cm respectively. The 7.07 and 8.13% of medullated fibres (PMF) in the present study was similar to that reported for two genotypes in second generation (Mokhber, 2005) and smaller than 9.42 and 18.36% in pure breeds of Ghezel and Moghani respectively (Shodja *et al.*, 2004). This amount in fine wool breeds is about 0.1 to 0.5%. Proportion of kemp fibre (PK) in ArGh and ArMo was 1.02 and 2.71% respectively. However, the large variation in proportion of kemp fibre, an undesirable trait demonstrates an opportunity for improving ArGh and ArMo fleece quality by selection. The amount of medullated fibres in the current study falls within the range of 5 to 25% reported for carpet-wool breeds in Asia (Mehta *et al.*, 2004).

The comfort factor is the percentage of fibres less than 30 micron and is determined when measuring the fibre diameter distribution of a sample using Projection Microscope, OFDA, etc. The importance of the comfort factor relates to the feel of a fabric on a wearer's skin. Fabric made from wool with a high comfort factor (>95%) will have less rigid fibres that bend more easily and therefore feel less 'prickly' (AWTA, 2008). In the present study the comfort factor for ArGh and ArMo was 68.46% and 63.26% respectively. Comfort factor has not been reported in carpet-wool breeds, so we were unable to compare these results with other literature. Nonetheless, it seems that the comfort factor is low in carpet-wool sheep breeds due to their high average fibre diameter.

The values presented in Table 3 are least square means of the untransformed data; the attached probabilities and tests relate to the transformed data. This procedure was followed, as back transformation of least square means of transformed values is mathematically inappropriate. The influence of fixed effects (genetic group, sex, birth type and year of birth) on wool traits was studied.

Average fibre diameter

A large variance is given in literature regarding the fixed effects influence on fibre diameter. In the present study genotype had a significant effect ($P<0.05$) on average fibre diameter and ArGh crossbreeds showed a significantly lower diameter than ArMo but the difference was only about 2 μm. Differences between genotypes were found by Bunge *et al.* (1996) and Lupton *et al.* (2004), but not by Yilmaz *et al.* (2003), and Mokhber (2005). One explanation for the differences between two genotypes seems to be the direct effects of Ghezel and Moghani pure breeds. Year of birth had a significant effect ($P<0.05$) on AFD. This result was in agreement with the findings of Iniguez *et al.* (1998) and Mokhber (2005). Animals which were born in 2007 had less AFD than animals that were born in 2008 (Table 3). There was no significant effect of sex on AFD. Tabbaa *et al.* (2001) and Mehta *et al.* (2004) reported similar results for the effect of sex. Birth type affected ($P<0.05$) AFD in two genotypes and single-born animals had more fibre diameter than twins. Mokhber (2005) reported a similar result for the effect of birth type on AFD in second generation crossbreeds.

Fibre diameter variability

The influence of fixed effects on FDV in F_3 of ArGh and ArMo was exactly the same as the results of Mokhber (2005) for F_2 of crossbreeds. Genotype did not affect FDV which is in agreement

Table 3. Least square means, standard errors and effect of factors affecting fleece characteristics according to genotype, sex, year of birth and birth type.

Subclass	#	AFD (μm)	FDV	SL (cm)	PK (%)	PMF (%)	CF (%)
Genotype		*	ns	ns	**	ns	ns
ArGh	154	27.29[a]±0.66	39.05±2.26	11.96±1.24	1.26[a]±0.49	2.23±0.25	68.93±2.66
ArMo	142	29.54[b]±0.91	36.09±1.99	11.56±1.27	2.44[b]±0.67	2.64±0.35	63.33±3.66
Sex		ns	**	ns	ns	**	ns
Male	141	27.90±0.77	35.25[a]±1.84	12.18±1.40	1.61±0.57	2.18[a]±0.29	67.22±3.08
Female	155	28.93±0.67	39.53[b]±2.2	11.49±1.25	2.34±0.54	2.78[b]±0.25	65.04±2.71
Year of birth		*	ns	**	ns	ns	ns
2007	114	27.41[a]±1.01	35.93±2.89	9.77[b]±1.37	2.38±0.74	2.41±0.38	68.01±4.06
2008	182	29.43[b]±0.50	38.48±1.46	12.15[a]±0.60	1.32±0.37	2.46±0.19	64.24±2.01
Birth type		*	*	**	ns	ns	ns
Single	180	28.71±0.50	39.43[b]±1.49	11.93[a]±1.25	2.39±0.37	2.52±0.19	64.81±2.01
Twins	116	28.12±1.02	35.72[a]±3.01	11.03[b]±1.42	1.31±0.75	2.35±0.39	67.45±4.10

AFD: average fibre diameter, FDV: fibre diameter variability, SL: staple length, PK: kemp fibres, PMF: proportion of medullated fibres, CF: comfort factor.
ns: non significant, *$P<0.05$, **$P<0.01$; in each of the subgroups, means followed by different superscripts differ significantly.

with the findings of Yilmaz et al. (2003). Sex ($P<0.01$) and birth type ($P< 0.05$) had a significant effect on FDV. McGregor and Butler (2004) did not find any significant effect of sex on FDV of Australian Alpacas. In contrast, Dashab et al. (2006) reported that sex had a significant effect on FDV in Naeini sheep. FDV for females was higher than for males and for singles was higher than for twins. The main reason for sex differences may be hormonal differences between two sexes and the genes located on the sex chromosomes. It seems that females are under more metabolic stresses than males (because of pregnancy and milk production) and therefore the availability of nutrients to wool follicles fluctuates. Thus, FDV would be higher for females. Research has shown that FDV increases exponentially with fibre diameter (Rafat et al., 2007). So, it could be concluded that higher FDV in single-born animals is the outcome of their high fibre diameter in contrast to twin-born animals.

Staple length

Genotype group did not have a significant effect on SL, which is in agreement with Bunge et al. (1996) and Shodja et al. (2004). No differences between sexes could be found for SL. These results correspond with some research but are contrary to other studies (Iniguez et al., 1998; Dick and Sumner, 1996; Dashab et al., 2006). SL was affected by year of birth and birth type ($P<0.01$); the significance of year of birth is in agreement with Lupton et al. (2004) and Mokhber (2005) and is contrary to Sidwell et al. (1973). Animals which were born in 2007 had shorter SL than those born in 2008 and this difference was significant ($P<0.01$). This difference was probably the result of the bad nutrition situation in 2008. As shown in Table 3, birth type had significant effects on SL ($P<0.05$) and singles had longer SL than twins. It is possible that poor nutrition (especially for dams with twin lambs) during early postnatal life imposes a permanent limitation on fibre production ability and fibre diameter of lamb (Dick and Sumner, 1996).

Proportion of kemp and medullated fibres

In the present study, kemp and medullated fibres are distinguished. Some authors give the proportion of coarse hair ('cerdas') and the diameter that goes with it (Wurzinger, 2005; Ayala, 1999 and Parra, 1999). Iniguez et al. (1998) defined kemp as proportion of fragmented and continuous medullated fibre. The results in Table 3 show a significant effect ($P<0.01$) of genotype on PK. In this study ArMo crossbreeds had higher PK than ArGh. This result is in agreement with the results reported by Tekin et al. (1998) and Çolakoglu and Ozbyaz (1999). Shodja et al. (2004) reported that Moghani breed sheep has a higher proportion of kemp than Ghezel, so it could be concluded that direct effect of Moghani breed has caused the differences between two genotype groups. Mokhber (2005) reported similar results for the effect of genotype on PK in the second generation. Effects of Sex, birth type and year of birth on PK were not significant. These results were in agreement with the findings of Iniguez et al. (1998) and Wurzinger et al. (2005).

No differences between genotype groups, year of birth and birth type could be found for medullated fibres (Table 3). However, the proportion of medullated fibres was higher in ArMo and single-born animals. These results are in agreement with Tabbaa et al. (2001) and Çolakoglu and Ozbyaz (1999) but contrary to other studies (Lupton et al., 2006; Çorekci and Evrum, 2000). Sex had a highly significant effect ($P<0.01$) on PMF, which is in agreement with the results reported by Lupton et al. (2004) on Dorset and Texel breed and Mokhber (2005) on ArGh and ArMo crossbreeds. Regarding Table 3, females had higher PMF than males (2.78% versus 2.18%). This result can be attributed to delay in follicle maturation in females because of their physiological conditions. PMF in singles was higher than twins but this difference was not significant ($P>0.05$). Regarding higher AFD in singles than twins, this result is logical and acceptable because, high positive correlation has been reported between AFD and PMF (Safari et al., 2005).

Comfort factor

Comparisons of the comfort factor should be made with care. In the present study, the definition is valid for proportions of fibres with a diameter less than 30 μm, Ayala (1999) and Parra (1999) define it as 'fibra descerdada' (de-haired fibre), whereas Tekin *et al.* (1998) define it as non-medullated fibre and Martinez *et al.* (1997) as a combination of unmedullated and fragmented medullated fibre. None of the studied effects had a significant influence on comfort factor (Table 3). This trait has not been studied in carpet-wool breeds, so it is hard to discuss the validity or invalidity of the given results. CF in ArGh crossbreeds, males and twins was higher than ArMo, females and singles respectively and these differences were not significant ($P>0.05$) but logical, because an increase in AFD and a simultaneous decrease in comfort factor have been reported. This result is in line with Parra (1999) and Martinez *et al.* (1997). In addition, a highly negative correlation has been reported among these two traits (Iniguez *et al.*, 1998). According to the results of this study, crossbreeding with Arkharmerino has shown favourable effects on wool characteristics improvement of Azerbaijan native breeds. Crossbreed sheep have shown to produce finer wool than native breeds with better uniformity. By conducting well planned designs, crossbreeds can supply the wool needed for fine carpet industries. Improvements in management and breeding strategies should be carried out at the same time.

References

AWTA (Australian Wool and Textile Association), 2008. Wool diameter: fact and fiction. Australian Wool and Textile Association. Available at: http://www.awta.com.au/ Publications/Fact Sheets/Fact Sheet 005.htm.

Ayala C., 1999. Caracter´ısticas f´ısicas de la fibra de llams j´ovenes. Progress in South American Camelids Research. EAAP Publication No. 105, pp. 27-34.

Bunge R., D.L. Thomas, T.G. Nash and C.J. Lupton, 1996. Performance of Hair Breeds and Prolific Wool Breeds of Sheep in Southern Illinois: Wool Production and Fleece Quality. Journal of Animal Science, 74: 25-30.

Cedillo R.M., W. Hohenboken and J. Drummond, 1977. Genetic and environmental effects on age at first estrus and on wool and lamb production of crossbred ewe lambs. Journal of Animal Science, 44: 948.

Çolakoglu N, and C. Özebyaz, 1999. Comparison of some production traits in Malya and Akkaraman sheep. Turkish Journal of Veterinary and Animal Sciences, 23: 351-360.

Çorekci S.G. and M. Evrum, 2000. Comparative studies on the production performances of Chios and Imroz breeds kept under semi-intensive conditionus II. Milk production, fleece yields and characteristics. Turkish Journal of Veterinary and Animal Sciences, 24: 545-552.

Dashab G., M.A. Edriss, A.A. Ghare Aghaji, H. Movasagh and M.A. Nilforooshan, 2006. Wool fiber quality of Naeini sheep. Pakistan Journal of Biological Sciences, 2: 270-276.

Delgado J., A. Valle Zarate and C. Mamani, 1999. Fiber quality of a Bolivian meat-oriented llama population. Progress in South American Camelids Research EAAP Publication, 105: 101-109.

Delgado J., 2003. Perspectivas de la producci´on de fibra de llama en Bolivia. Dissertation, Universität Hohenheim, Germany.

Demir H. and H. Baspinar, 1992. Production performance of Kivircik sheep under semi-intensive conditions. 2. Fertility, milk production, body weight and wool characteristics of ewes. Journal of Faculty of Veterinary Medicine University, 17: 13-24.

Dick J.L. and R.M.W. Sumner, 1996. Development of fiber and follicle characteristics related to wool bulk in Perendale sheep over the first year of life. Proceedings of the New Zealand Society of Animal Production, 56: 314-318.

Fahmy M.H., 1987. The accumulative effect of Finnsheep breeding in crossbreeding schemes: wool production and fleece characteristics. Canadian Journal of Animal Science, 67: 1-11.

Iniguez L.C., R. Alem, A. Wauer and J. Mueller, 1998. Fleece types, fiber characteristics and production system of an outstanding llama population from Southern Bolivia. Small Ruminant Research, 30: 57-65.

Kempester A.J., D. Croston, D.R. Guy and D.W. Jones, 1987. Growth and carcass characteristics of crossbred lambs by ten sire breed, compared at the same estimated carcass subcutaneous fat proportion. Animal Production, 44: 83-98.

Lupton C.J., B.A. Freking and K.A. Leymaster, 2004. Evaluation of Dorset, Finnsheep, Romanov, Texel, and Montadale breeds of sheep: III. Wool characteristics of F_1 ewes. Journal of Animal Science, 82: 2293-2300.

Lupton C.J., A. McColl and R.H. Stobart, 2006. Fiber characteristics of the Huacaya Alpaca. Small Ruminant Research, 64: 211-224.

Martinez Z., L.C. Iniguez and T. Rodriguez, 1997. Influence of effects on quality traits and relationships between traits of the llama fleece. Small Ruminant Research, 24: 203-212.

McGregor B.A. and K.L. Butler, 2004. Sources of variation in fiber diameter attributes of Australian alpacas and implication for fleece evaluation and animal selection. Australian Journal of Agricultural Research, 55: 433-442.

Mehta S.C., S.K. Choprn, V.K. Singh, M. Ayub and V. Mahrotra, 2004. Production and quality of wool in Magra breed of sheep. Indian Journal of Animal Science, 74: 792-794.

Mokhber M., 2005. Evaluation of wool characteristics of Arkharmerino × Ghezel and Arkharmerino × Moghani crossbreed sheep in second generation. MSc Thesis, University of Tabriz, Iran.

Parra G., 1999. Evaluacion del potencial productivo de la llama en la quinta seccion municipal charana. Tesis Ing. Agr., Universidad Mayor de San Simon, Cochabamba, Bolivia.

Rafat S.A., H. de Rochambeau, M. Brims, R.G. Thébault, S. Deretz, M. Bonnet and D. Allain, 2007. Characteristics of Angora rabbit fiber using optical fiber diameter analyzer. Journal of Animal Science, 85: 3116-3122.

SAS Institute Inc., 2004. SAS/STAT Software Release 9.1.SAS Institute Inc., Cary, NC, USA.

Safari E., N.M. Fogarty and A.R. Gilmour, 2005. A review of genetic parameter estimates for wool, growth, meat and reproduction traits in sheep. Livestock Production Science, 92: 271-289.

Shodja J., T. Farahvash and S.A. Rafat, 2004. The evaluation of fleece characteristics in Arkharmerino×Moghani F_1 crossbreds. Proceedings of the 11th AAAP Animal Science Congress. 5-9th September. Kuala Lumpour, Malaysia, pp. 624-626.

Sidwell G.M., R.L. Wilson and M.E. Hourihan, 1973. Production in some pure breeds of sheep and their crosses. IV: effect of crossbreeding on wool production. Journal of Animal Science, 32: 1099-1102.

Tabbaa M.J., W.A. Al-Azzawi and D. Campbell, 2001. Variation in fleece characteristics of Awassi sheep at different ages. Small Ruminant Research, 41: 95-100.

Tekin M.E., M. Gurkan and R. Kadak, 1998. The wool characteristics of Akkaraman, Awassi and their F_1 and B_1 crossbreed with German Blackheaded Mutton and Hampshire Down breed. Animal Breeding and Genetics, 118: 92-94.

Wolf B.T., C. Smith and D.I. Sales, 1980. Growth and carcass composition in the crossbred progeny of six terminal sire breeds of sheep. Animal Production, 31: 307-313.

Wurzinger M., J. Delgado, M. Nurnberg, A. Valle Zarate, A. Stemmer, G. Ugarte and J. Solkner, 2005. Genetic and non-genetic factors influencing fiber quality of Bolivian llamas. Small Ruminant Research, 61: 131-139.

Yilmaz A., M. Ozcan, B. Ekiz, A. Ceyhan and A. Altinel, 2003. The body weights and wool characteristics of the indigenous Imroz and Kivircik sheep breeds of Turkey. Wool technology and sheep breeding, 51: 16-23.

Differences in fibre diameter profile between shearing periods in white Huacaya Alpacas (*Vicugna pacos*)

P. Mayhua[1], E.C. Quispe[1], M. Montes[2] and L. Alfonso[2]
[1]*Programa de Mejora de Camélidos Sudamericanos, E.A.P. de Zootecnia, FCI, Universidad Nacional de Huancavelica (UNH)- Huancavelica-Perú; edgarquispe62@yahoo.com*
[2]*Área de Producción Animal, Universidad Pública de Navarra, Spain*

Abstract

A total of 153 midside fibre samples, one per animal at left side, were obtained in two shearing periods (I Period: December 2009, n=109; II Period: March 2010, n=44) from white Huacaya alpacas belonging to CIPCS – Lachocc herd of National University of Huancavelica. Fibre diameter profiles (FDP) were obtained using the OFDA 2000 instrument, measuring fibre diameter every 5 mm along the entire length of staples (Brims *et al.*, 1999), in a subsample of between 400-800 fibres per animal. Different fibre diameter profile variables were calculated and compared between shearing periods taking into account differences due to sex and age. The variables were the following: average fibre diameter (AFD), maximum and minimum fibre diameter (FDMax, FDMin), its respective positions (PMax, PMin), fibre diameter change along the staple (TACA, calculated as the absolute value of (FDMax - FDMin), and fine ends (FE; calculated as the difference between the mean diameter of fibres in the first (f) and last (l) 5 mm of the staple, and FD: FE=[(f+l)/2]-FD). Results show different FDP for the two shearing periods, being significantly different PMax PMin and FE ($P<0.05$). No differences for sex were observed, but age affected AF, FDMax, FDMin, PMax, PMin, TACA and FE ($P<0.05$). These results showed, in terms of the needs of textile industry in processing performance, a better quality of fibre obtained during the shearing of December 2009.

Keywords: alpaca, fibre diameter, shearing

Introduction

The global population of alpacas is estimated at 3.7 million (FAO, 2005), and approximately 80% of them are in the high Andes in Peru (above 3,000 m). About 86% of the population are white alpacas (Brenes *et al.*, 2001). The Huacaya breed predominates in the country with 93%, while the Suri race is only 7% (FAO, 2005), although in some regions of Peru (as in the case of Huancavelica), the Huacaya breed can reach 96% (Oria *et al.*, 2009).

In the high-Andean regions, alpaca breeding is the main activity for the people living there as well as their main source of direct income. Furthermore, this animal produces a fleece with fibres that are highly valued and the garments made are classified as luxury items (Wang *et al.*, 2003) because of their unique physical characteristics. The knowledge of the production of alpaca fibre could improve the technological characteristics of the fibre textiles and promote better marketing of farmers in the textile industry, which would ensure better processing and improved production inputs (Quispe, 2010).

There are many characteristics that determine the quality of the fibre, but AFD (average fibre diameter) may have the greatest influence on price of the fleece and best use (McGregor, 2006). This parameter differs significantly between fleeces, body region, fibres within a sample, and also along the fibre, with varying percentages of 16, 4, 64 and 16, respectively within a herd (Hansford, 1992). Variation along the fibre, which helps form the fibre diameter profile (FDP) for a period of growth (usually one year), is caused by genetic effects (Brown *et al.*, 2002), physiological states [such as nursing and pregnancy (Hansford, 1992)], weather (season, photoperiod, temperature) and availability of pasture (Naylor and Stanton, 1997; Naylor and Hansford, 1999; and Naylor *et al.*, 2004). This last factor has a greater effect on the FDP, as is noted in sheep by Smith *et al.* (2006) and in alpacas by McGregor (1999) and Quispe *et al.* (2008a); it has been observed that malnutrition during certain times of the year has a major effect on the production and quality of alpaca fibre (Russel, and Redden, 1997; Franco *et al.*, 2009).

FDP is important for determining the quality, due to the effect on the appearance, comfort and product performance of fibre for textile processing (Mayo *et al.*, 1994). Therefore, further research is necessary if quality is to be improved. A FDP with fibre terminals ('fine ends') thinner than the middle (convex shaped) will not only be more comfortable (Naylor *et al.*, 2004), but also have greater tensile strength, which will give better performance during the textile process (Thompson and Hynd, 1998). On the other hand, the results of Hansford and Kennedy (1998) support the concept that abrupt changes in fibre diameter are associated with lower tensile strength (TS). Naylor and Hansford (1999) report differences between maximum and minimum diameter of between 5 to 12 microns.

Currently in the case of sheep, the FDP data helps producers manage their animals to produce wool with less variation throughout the year, leading to the production of finer and stronger wool that improves throughput (Peterson and Oldham, 2000). This same strategy could be followed, based on in-depth knowledge of the characteristics that define FDP in alpacas.

In South America, alpacas are raised strictly under an extensive system, with a diet based on natural pastures located in areas with rugged geography and harsh climate (Brenes *et al.*, 2001), so predictable changes along the diameter fibre are to be expected, as in the case of wool FDP, reflecting environmental changes occurring throughout the period of growth since the previous shearing (Brown *et al.*, 2002). There is little knowledge about alpaca FDP, with only a preliminary report in Peru by Quispe *et al.* (2008b). For this and the previously discussed reasons, we have conducted this research with the aim of characterising the FDP at different times of shearing, corresponding to two different periods of growth.

In particular, we propose to characterise the FDP of the white Huacaya alpaca in two periods of shearing, according to the characteristics of textiles, such as average fibre diameter (AFD), maximum and minimum fibre diameter (FDMax, FDMin), its respective positions (PMax, PMin), fibre diameter change along the staple (TACA) and fine ends (FE), obtained in two shearing periods under the environmental conditions and farming system prevailing in Huancavelica.

Materials and methods

Location

The field work was conducted at the Centre for Research, Production and Services 'Lachocc (CIPS-Lachocc) at to the National University of Huancavelica (UNH), located 35 km from Huancavelica, at an altitude of 4,250 m, with temperatures ranging from -5 °C to 0 °C at night and 14 to 18 °C during the day; annual precipitation reaches 790 mm per year.

Samples

Midside fibre samples (5 g), one per animal at left side, were obtained from 153 Huacaya white alpacas of different ages (75 males and 78 females), considering two annual periods of growth from two shearing periods [Period I: growth in December 2008 to December 2009 (n=109) and Period II: growth since March 2009-March 2010 (n=44)].Animal identity, sex, staple length, and fibre growth period were recorded.

Grazing animals

The alpacas were reared exclusively on pasture, consisting mainly of desirable species, such as *Alchemilla pinnata*, *Festuca dolichophylla* and *Distichia muscoid*; and undesirable species such as: *Alquigluma Poa*, *Calamagrostis ovata* and *Calamagrostis curvula*; and undesirable: *Aciachne pulvinata*, *Plantago rigida* and *Stipa obtusa*. Raising and health conditions were similar for all animals.

Sample analysis

Samples were evaluated by the optical fibre diameter analyser 2000 (OFDA 2000) in the UNH's Wool and Fiber Laboratory following Brims *et al.* (1999). OFDA 2000 performs measurements of staples from the tip to the base every 5 mm, which allows the generation of fibre diameter profile (FDP) of each sample. Because each sample had a different number of profile measurements, we proceeded to the standardisation ranging from 5%, and then to the characterisation of the profiles based on average fibre diameter (AFD), maximum and minimum fibre diameter (FDMax, FDMin), its respective positions (PMax, PMin), fibre diameter change along the staple (TACA) and fine ends (FE). FE was calculated as the difference between the mean diameter of fibres in the first (f) and last (l) 5 mm of the staple, and FD: FE=[(f+l)/2]-FD).

Results and discussion

FDP characteristics in two shearing periods

Table 1 shows that of seven variables that characterise the FDP, only PMin, PMax and FE are strongly influenced by shearing periods. It may also indicate that when the shearing is done in December (Period I) PMin is closer to the tip of the fibre, PMax is farther from the tip of the fibre, and FE is negative, results very different from shearing Period II (March). These results result in a convex FDP for Period I, which would be due to increased forage supply immediately after the previous shearing, with a decrease from the middle of the fibre, an effect contrary to that when shearing takes place in March, the month that coincides with the decline of rainfall in a short time prior to shearing, with a consequent decrease in pasture production. These results are consistent with those found in sheep, by Brown *et al.* (2002), Naylor and Stanton (1997), Naylor and Hansford (1999) and Smith *et al.* (2006), and would lead us to induce that shearing fleeces obtained in December would provide a better quality of fibre compared to those obtained in shearing March.

For AFD, FDMax and FDMin no significant differences were observed between the two periods of shearing. On the other hand, as shown in Table 2, while sex has no effect on any of the variables under study, age has a clear effect on all of them (AFD, PMin, PMax, FDMin, FDMax, TACA and FE).

Values are quite similar to those reported by Quispe *et al.* (2008b) and Poma *et al.* (2009), whose results for FDMin and FDMax were 21.8 and 24.7 μm, and of 20.1 and 25.2 microns respectively. The 5 μm of TACA are similar to values indicated by Curtis in 1982 (quoted by Naylor and Hansford, 1999), who estimates that there is a 5 to 6 μm exchange for wool from sheep in Australia. However, Naylor and Stanton (1997) indicate that these changes are detrimental at the

Table 1. Average ± SE and confidence interval of average fibre diameter (AFD), minimum and maximum fibre diameter position (PMin and PMax, respectively), minimum and maximum fibre diameter (FDMin and FDMax, respectively), fibre diameter change along the staple (TACA) and Fine ends (FE), in two shearing periods.

Variables	Periods				
	Period I: December		Period II: March		
	Average±SE	Confidence interval	Average±SE	Confidence interval	
AFD (μm)	20.95±0.44	20.07-21.83	20.82±0.48	19.86-21.76	
PMín (%)	38.65±5.99	26.81-50.49	63.12±6.46	50.36-75.89	
PMax (%)	55.28±4.49	46.39-64.17	25.75±4.85	16.17-35.33	
FDMin (μm)	18.70±0.41	17.90-19.51	18.81±0.44	17.94-19.67	
FDMax (μm)	23.84±0.52	22.80-24.87	23.89±0.56	22.77-25.00	
TACA (μm)	5.13±0.29	4.55-5.72	5.08±0.32	4.45-5.71	
FE (μm)	-0.92±0.28	(-1.48)-(-0.36)	1.36±0.31	0.76-1.96	

Table 2. Least square means for average fibre diameter (AFD), minimum and maximum fibre diameter position (PMin and PMax, respectively), minimum and maximum fibre diameter (FDMin and FDMax, respectively), fibre diameter change along the staple (TACA) and fine ends (FE), per period, sex and age factors and period × age interactions. Significance of factors is also indicated.

Factors	AFD (μm)	PMin (%)	PMax (%)	FDMin (μm)	FDMax (μm)	TACA (μm)	FE (μm)
Shearing periods	n.s.	*	**	n.s.	n.s.	n.s.	**
I: December	20.95	38.65	55.28	18.70	23.84	5.13	-0.92
II: March	20.82	63.12	25.75	18.81	23.89	5.08	1.36
Sex	n.s.	n.s.	n.s.	n.s.	n.s.	n.s.	n.s.
Male	20.47	52.12	41.89	18.38	23.51	4.97	0.30
Female	21.30	49.66	39.14	19.13	24.37	5.24	0.14
Age	**	**	**	**	**	**	*
< 1.5 years old	20.07	64.59	37.58	18.58	22.44	3.86	-1.02
1.5 a 3 years old	20.10	41.33	34.67	17.79	22.86	5.07	0.52
3 a 4 years old	20.68	36.74	67.35	17.90	23.69	5.79	-1.00
> 4 years old	23.41	53.57	33.93	21.28	26.62	5.4	1.20
Period × Age	**	n.s.	n.s.	**	**	n.s.	**
Period I ×							
< 1.5 years old	20.53	62.72	41.30	19.01	22.80	3.80	-1.42
1.5 a 3 years old	19.45	22.10	52.80	17.24	22.04	4.80	-1.91
3 a 4 years old	20.51	28.00	74.40	17.57	23.40	5.83	-1.57
> 4 years old	23.31	41.77	52.63	20.99	27.09	6.10	1.22
Period II ×							
< 1.5 years old	17.06	75.87	15.66	15.78	19.96	4.18	1.45
1.5 a 3 years old	20.39	54.56	23.04	18.04	23.24	5.20	2.16
3 a 4 years old	22.16	62.47	41.37	19.80	25.77	5.96	0.67
> 4 years old	23.64	59.58	22.93	21.60	26.59	4.99	1.16

* $P<0.05$; ** $P<0.01$; n.s.= not significant.

time of marketing. Major changes to 6 μm in sheep are due to the effects of physiological stress (pregnancy and lactation) and greater than 8 μm are due to reduction in the quantity and quality of food (Hansford, 1992), which will significantly reduce the tensile strength, due to the smaller amount of keratinic material that will withstand breaking force. This result shows that alpacas stressors apparently have an effect on the diameter, which would strengthen the hypothesis of the great adaptability to harsh conditions that these animals have acquired (Quispe *et al.*, 2008a).

Fibre diameter profile in the two shearing periods

FDP for both periods are shown in Figure 1, where it can be seen that the fibres obtained in Period I have an FDP convex. Meanwhile, FDP obtained for the Period II is concave, showing a radical difference, which may be for two main reasons: (1) an increase in forage availability during the months from December to April, which corresponds to distances between 40 and 70% of the period I, and distances distributed between 0-10% and between 80-90% for the period II, and (2) the effect of physiological status (pregnancy and lactation); it should be noted that most of the animals sheared in the period II (March) are females who recently gave birth and because of this nutrients were previously directed to the maintenance of the foetus, which, coupled with low forage availability, would give the profile of this period a fall position of 60 to 90%.

Conclusions

It can be concluded that fleeces obtained during shearing Period I (December) would have a better quality than those obtained in shearing Period II (March). The first would have better comfort and resistance, because they have negative FE, which leads to better performance during the textile process, better traction to resist those subjected to scouring, washing and combing. Therefore, disadvantages of the fleeces collected in the Period II (March), could be overcome by changing the date of shearing, but also through appropriate dietary supplementation of pregnant females.

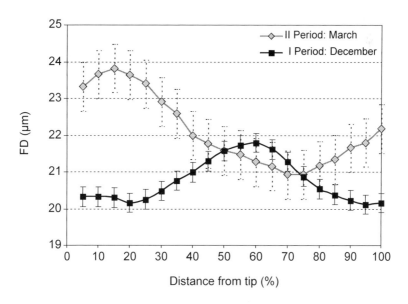

Figure 1. Fibre diameter profiles in two shearing periods: I and II, which corresponds to shearing in December and March, respectively.

References

Brenes E.R., K. Madrigal, F. Pérez and K. Valladares, 2001. El Cluster de los Camélidos en Perú: Diagnóstico Competitivo y Recomendaciones Estratégicas. Instituto Centroamericano de Administración de Empresas. Available at: http://www.caf.com/attach/4/default/CamelidosPeru.pdf. Accessed 21 March 2009.

Brims M.A., A.D. Peterson and S.G. Gherardi, 1999. Introducing the OFDA2000 - For rapid measurement of diameter profile on greasy wool staples. IWTO, Raw wool group report RWG04, Florence, Italy.

Brown D.J., B.J. Crook and I.W. Purvis, 2002. Differences in fibre diameter profile characteristics in wool staples from Merino sheep and their relationship with staple strength between years, environments and bloodlines. Australian Journal of Agricultural Research, 53: 481-491.

FAO (Food and Agriculture Organization), 2005. Situación Actual de los Camélidos Sudamericanos en el Perú. Organización de las Naciones Unidas para la Agricultura y la Alimentación. Proyecto de Cooperación Técnica en apoyo a la crianza y aprovechamiento de los Camélidos Sudamericanos en la Región Andina TCP/RLA/2914. Available at: http://www.rlc.fao.org/es/ganaderia/pdf/2914per.pdf. Accessed 10 September 2009.

Franco F., F. San Martín, M. Ara, L. Olazábal and F. Carcelén, 2009. Efecto del nivel alimenticio sobre el rendimiento y calidad de fibra en alpcas. Revista de Investigaciones Veterinarias del Perú, 20: 187-195.

Hansford K.A., 1992. Fibre Diameter Distribution: Implications for Wool Production. Wool Technology and Sheep Breeding, 40: 2-9.

Hansford K.A. and J.P. Kennedy, 1998. Relationship between the rate of change in fibre diameter and staple strength. Proceedings of the Australian Society of Animal Production, 17: 415.

Oria I., I. Quicaño, E. Quispe and L. Alfonso, 2009. Variabilidad del color de la fibra de alpaca en la zona altoandina de Huancavelica-Perú. Animal Genetic Resources Information, 45: 79-84.

Mayo O., B. Crook, J. Lax, A. Swan and T.W. Hancock, 1994. The determination of Fibre Diameter Distribution. Wool Technology and Sheep Breeding, 42: 231-236.

McGregor B.A., 1999. The influence of environment, nutrition and Management on the quality and production of alpaca fibre. Proceedings: Alpaca – Born to be Worn, Australian Alpaca Industry Conference, Australia, pp. 88-94.

McGregor B.A., 2006. Production attributes and relative value of alpaca fleeces in southern Australia and implications for industry development. Small Ruminant Research, 61: 93-111.

Naylor G.R.S. and J. Stanton, 1997. Time of shearing and the diameter characteristics of fibre ends in the processed top: An opportunity for improved skin comfort in garments. Wool Technology and Sheep Breeding, 45: 243-255.

Naylor G.R.S. and Hansford K.A., 1999. Fibre End Diameter Properties in Processed top Relative to the Staple for Wool Grown in a Meditarranean Climate and Shorn in Different Seasons. Wool Tech. Sheep Breeding, 42: 107-117.

Naylor G.R.S., C.M. Oldham and J. Stanton, 2004. Shearing time of mediterranean wools and fabric skin comfort. Textile Research Journal, 74: 322-328.

Peterson A.D., S.G. Gherardi and M.R. Ellis, 2000. Managing the diameter profile leads to increased staple strength of young Merino sheep shorn in spring in South Western Australia. The Asian-Australian Journal of Animal Science, 23: 469-472.

Poma A.G., C.E. Ventura and E.C. Quispe, 2009. Caracterización del perfil del diámetro de fibra en alpacas Huacaya de color blanco. Thesis, Universidad Nacional de Huancavelica, Peru, 99 p.

Quispe E.C., J.P. Mueller, J. Ruiz, L. Alfonso and G. Gutiérrez, 2008a. Actualidades sobre adaptación, producción, reproducción y mejora genética en camélidos. Universidad Nacional de Huancavelica. Primera Edición. Huancavelica, Perú, pp. 93-112.

Quispe E.C., R. Paúcar, A. Poma, D. Sacchero and J.P. Mueller, 2008b. Perfil del diámetro de fibras en alpacas. Proc. de Seminario Internacional de Biotecnología Aplicada en Camélidos Sudamericanos. Huancavelica. Perú.

Quispe E.C., 2010. Estimación del progreso genético de seis esquemas de selección en alpacas (Vicugna pacos) Huacaya con tres modelos de evaluación en la región alto andina de Huancavelica. Thesis. UNALM. 167 p.

Russel A.J. and H.L. Redden, 1997. The effect of nutrition on fibre growth in the alpaca. Animal Science, 64: 509-512.

Smith J.L., I.W. Purvis and G.J. Lee, 2006. Fibre diameter profiles – potential applications for improving fine-wool quality. International Journal of Sheep and Wool Science, 54: 54-61.

Thompson A.N. and P.I. Hynd, 1998. Wool growth and fibre diameter changes in young Merino sheep genetically different in staple strength and fed different levels of nutrition. Australian Journal of Agricultural Research, 49: 889-898.

Wang X., Wang L. and X. Liu, 2003. The Quality and Processing Performance of Alpaca Fibres. RIRDC publication N° 03/128: 66-76.

Fibre production and fibre characteristics of alpacas farmed in United States

T. Wuliji
Department of Animal Biotechnology, University of Nevada, Reno, NV 89557, USA; Present
address: Department of Agriculture and Environmental Sciences, Lincoln University, Jefferson
City, MO 65102, USA; tumenw@yahoo.com

Abstract

The alpaca is the most important fibre-producing member of the South American camelids. There are two types of alpacas introduced into the United States, namely, Huacaya and Suri, however, a majority of alpacas are that of the Huacaya breed. Currently, there are 171,316 alpacas registered in the Alpaca Registry Inc (ARI) from 1986 to 2010 in the U.S. Alpacas can be found in every state of the United States and are farmed in various geographical environments ranging from hot desert to high mountain ranges. Alpacas were shorn at 10 to 18 months of fibre growth intervals and produced 2 kg per head fleece per year. Coat colour is varies widely in the alpacas, ranging from white to black and various shade combinations in 22 different natural colour categories. An earlier survey in different U.S. regions for Huacaya alpaca performance recordings showed that body weight, fleece weight, and fibre quality traits have improved for alpacas farmed in North America. This paper presents the recent analysis of both Huacaya (n=714) and Suri (n=502) alpacas sampled at 18 alpaca ranches located within the west, central and eastern regions of the U.S. Body weight, average fibre diameter, fibre diameter variation, mean staple length, and comfort factor estimates were 61.8 kg, 24.9 micron, 19.4%, 74.5 mm and 81.4% for Huacaya and 65.5 kg, 26.5 micron, 20.7%, 75.5 mm and 77% for Suris. Although it appeared that Suri alpacas were heavier for body weight and about 1.5 micron coarser than Huacaya fleeces tested in this study, there was no evidence for any fibre production or fibre characteristic superiority in the one breed over the other except the preference of a breed specialty trait.

Keywords: alpaca, fleece weight, fibre diameter, coat colour

Introduction

Camelids are indigenous to the Andean highlands of South America. The alpaca (*Lama pacos*) is the most numerous fibre-producing member of the four South American camelid species of Llama, guanaco, alpaca and vicuna (Kadwell *et al.*, 2001). There are two types of alpacas introduced into the United States, namely, Huacaya and Suri, however, the majority of alpacas (80%) are that of the Huacaya breed. Although almost physically identical, what distinguishes the two breed of alpacas is their fleece type. Huacaya fleece is characterised as short, dense, crimpy and woolly compared to the Suri, which is described as silky, lustrous, long and with pencil locks (Wuliji *et al.*, 2000). While Peru owns 75% of alpaca population and produces about 90% of the world camelid fibre (Morante *et al.*, 2009), there is a steady increase in the alpaca population in the introduced regions, such as Australia, Canada, Great Britain, New Zealand, and the United States, and it is becoming an important livestock species. There are 171,316 alpacas registered to the Alpaca Registry Inc (ARI) from 1986 to 2010 in the U.S., and the total population of camelids has been estimated to exceed the ¼ million head count. Now, alpacas can be found in every state of the United States and are farmed in various geographical environments ranging from the hot desert to high mountain ranges. Alpacas were shorn at 10 to 18 month of fibre growth intervals and produced about 2 kg of fleece per head. Coat colour varies widely in the alpacas, ranging from white to black and various shades differing in 22 natural colour categories. Alpaca fibre production traits have been reported in several studies in their native lands (Pumayalla and Leyva, 1988; Frank *et al.*, 2006; Morante *et al.*, 2009; Cervantes *et al.*, 2010) as well as in the introduced environments (Wuliji *et al.*, 2000; McGregor and Butler, 2002; McColl *et al.*, 2004). The profiling of fibre characteristics in Huacaya breed in the U.S. farmed alpacas was reported (Lupton *et al.*, 2006). The current study was conducted to compare production performance measured for body weight, fleece weight, fleece blanket weight, and some major fibre characteristics in two breeds of alpacas – Huacaya and Suri in the U.S. registered herds.

Materials and methods

Alpacas and data collection

This investigation was initiated at the request of the members of the alpaca breeders' associations in the western U.S. region, who expressed a strong interest in accessing comparable alpaca fibre production performance and fibre quality characteristic information on Huacaya and Suri alpacas, which can be referenced for their breeding programmes. Forty alpaca producer members of the Alpaca Owners and Breeders Association (Inc), distributed across in the western, central and eastern geographical regions of the U.S. were randomly selected and invited to participate on a voluntary basis in the study. Participants are requested to provide their alpaca herd data including alpaca registration identification, names, animal performance records, age, coat colour and side fleece samples. No comparison was made available between or among ranches. An invitation was mailed to all selected members in May 2010. However, some producers had already either completed their shearing operation, or had no performance records, or had no fleece sampling, so were unable to contribute analytical data sets. The result animal performance recordings from 1,216 alpacas (Huacaya=714 and Suri=502) were provided by 18 participant ranchers. These on-farm recorded production data and in-lab fibre measurements were included for the statistical analysis. Alpaca numbers, production data, breed, and herd size for fleece sampling varied from ranch to ranch, ranging from a smaller group of 10 to the larger group of 250 alpacas. The majority of participants have provided either Suri or Huacaya alpaca data or fibre samples while a few producers included both breeds of alpacas. On-ranch records and fleece characteristic definitions were accepted and used as owner's provision, such as Alpaca Registry ID, animal names, body weight (some recorded as post-shearing body weight), fleece weight or fleece blanket weight, sex, age, and coat colours. However, some regroupings were used for the purpose of this analysis, for example, dark brown, brown and light brown were pooled as brown; rose gray, silver gray and gray were grouped as gray; and 'gelded male', 'breeding male' and 'non-breeding male' were pooled into male group, while animals over 8 years old were pooled into 8+ age group, which included alpacas from 8 to 20 years old.

Fleece and fibre measurements

A side fleece sample was obtained from alpacas at shearing with fleece weight or fleece blanket weight recorded. The sample size of 5-10 grams per fleece or fleece blanket was packed individually in the labelled plastic bags. Ranches were requested to mail the fleece samples with a field data recording sheet in a postal parcel addressed to Wool Testing Lab, University of Nevada, Reno. However, some fleece samples were provided without body weight, fleece weight or fleece blanket weight information. Upon reception, alpaca fibre samples were sorted for each ranch owner, and prepared for a set of fibre characteristic measurements using OFDA 2000 equipment (IWG, Australia). The OFDA 2000 equipment was programmed to measure a set of 25 characteristic measurements including average fibre diameter (AFD), fibre diameter standard deviation (FSD), diameter coefficient of variation (FCV%), spinning fineness fibre diameter (SF), fibre diameter variation along the length (FSDL), spinning fineness (SF), staple length (SL), Hauteur length (HL), comfort factor estimate (CF%), fibre curvature (CRV), curvature variation (CRVD), minimum fibre diameter (MFD), maximum fibre diameter (XFD), and fibre bulk (NZ Bulk) trait. Fibre measurement samples were prepared in accordance with the OFDA 2000 operating procedures, except that the grease correction factor (0.3-0.4%) was set specifically for alpaca fibre by measuring 30 scoured and non-scoured alpaca fleece samples for each ranch. The SL was estimated from the fibre length measured on OFDA 2000 equipment, so that the weathered staple tips and the staple base portion longer than the slide length were excluded. Upon completion of fibre tests, fibre measurement data, ranch summary statistics and individual fibre diameter histograms were provided to each ranch owner for review or verification respectively.

Data analysis

The statistical procedures of SAS (GLM, ANOVA, CORR and TTEST) were used for data analysis and interpretation for body weight, fleece weight, age, coat colour and fibre characteristic measurements. The main effects are primarily compared for breed, sex, age, and coat colours.

Results and discussion

BWT, FWT and fibre characteristics are presented by breed in Table 1. The BWT, AFD, FSD, SF and CV were significantly higher in Suris than Huacayas, although the numerical scale of difference was small. Huacaya alpacas had higher FWT, BFW, CF, MFD, XFD, NZ Bulk, CRV, and CRVD than Suri breed, whereas there was no difference in FSDL, SL, and age structure of the two breeds. The geographical region difference was not evaluated in this study due to the smaller number of ranches representing the regions. However, animals from the west coast and central region were found to be lighter in weight and finer in fleece for the Huacaya breed in the earlier survey (McColl *et al.*, 2004).

The AFD measured in this study appeared to be similar to or slightly coarser than alpacas reported in the Peruvian studies (Morante *et al.*, 2009; Cervantea *et al.*, 2010) but was finer than the introduced alpacas reported elsewhere (Wuliji *et al.*, 2000; McGregor and Butler, 2004). Although there was no difference in SL between Huacaya and Suri, the visual appearance of Suri fleeces showed they were longer than those of Huacayas. A crimpless long staple with pencil locks are of this breed's signature fibre type. A recent investigation of U.S. born alpacas indicated a single autosomal dominant gene controls Suri fleece type, which when subjected to suppression by modifier was segregating within Huacaya × Suri progeny (Sponenburg, 2010). The fleece weight and major fibre characteristics heritability were estimated to be moderate to high (Wuliji *et al.*, 2000; Frank *et al.*, 2006; Cervantes *et al.*, 2010). Washing yield, medullation, and staple strength were not determined in this study but these proved to be exceptionally high in fibres of Huacaya alpacas (yield=90%, staple strength= >30N/ktex, 60%) in the United States (McColl *et al.*, 2004).

Significant difference between sexes was found in BWT, FWT, FBT, CV and age structures (Table 2). Lupton *et al.* (2006) found a significant effect of sex, age and region and coat colour on the measured traits. Similarly, the age, breed and sex had exerted a significant effect on body weight, fleece weight and some fibre characteristics in this study. Males were found to produce

Table I. Body weight, fleece weight, fleece blanket weight and fibre characteristics measured in Huacaya (n=714) and Suri (n=502) alpacas.

Breed	Huacaya			Suri		
Traits	Mean	SD	n	Mean	SD	n
BWT (kg)	61.8	15.9	104	65.5[*]	16.7	382
FWT (kg)	2.5	1.1	212	2.2[**]	0.8	388
FBW (kg)	2.9	1.3	211	1.2[**]	0.4	377
AFD (μm)	24.9	4.9	713	26.5[**]	4.4	471
FDSD (μm)	4.8	1.2	713	5.4[**]	1.2	471
CV (%)	19.4	2.8	713	20.7[**]	3.0	471
SF (μm)	23.7	5.1	422	25.9[**]	4.4	449
SL (mm)	74.5	28.3	422	75.5	38.5	449
CF (%)	81.4	22.0	422	76.9[**]	21.1	449
MFD (μm)	21.9	7.4	422	18.3[**]	12.4	449
XFD (μm)	25.4	7.0	422	21.2[**]	12.6	449
FSDL (μm)	1.4	2.7	422	1.5	4.3	449
Bulk (cm³/g)	20.5	1.4	421	16.5[**]	0.5	449
HL (mm)	71.4	25.8	421	71.5	37.6	442
CRV (°/mm)	33.4	9.2	422	10.8[**]	3.1	449
CRVD°/mm)	26.2	6.3	422	12.0[**]	4.0	449
AGE (yr)	4.8	3.6	413	4.5	3.5	447

[*] $P< 0.05$; [**] $P< 0.01$.
BWT: body weight; FWT: fleece weight; AFD: average fibre diameter; FBT: fleece blanket weight; FSD: fibre diameter standard deviation; CV: coefficient of variation; CF: comfort factor; SF: spinning fineness; SL: staple length; MFD: minimum fibre diameter; XFD: maximum fibre diameter; FSDL: fibre diameter variation along the length; Bulk: New Zealand bulk measurement; HL: Hauteur length; CRV: fibre curvature; CRVD: curvature variation.

slightly heavier fleece weight with lighter BWT than females. However, there was no difference in AFD and other fibre measurements. Body weight, fleece weights and AFD increased with increasing age groups (Table 3). BWT increased significantly from age 1 to 6 and then gradually decreased in 7 and 8+ groups, however, AFD was continuously coarser at about 1 micron yearly from one year of age to 8 year olds. Body weight, fleece weight and fibre characteristics varied among coat colour groups (Table 4), where black, beige and light fawn groups showed a heavier FWT than other groups with slightly coarser fleeces. The grey group showed the lowest body weight and fleece weight recordings but a coarser AFD in this analysis. Other characteristic differences by coat colour were small and negligible. Overall, differences attributable to coat colour were found to be small in alpacas (Lupton *et al.*, 2006).

Pearson's correlation coefficients in the selected number of traits are given in Table 5. There were a few numerically high and significant correlations, such as, between BWT and AFD, CRV and Bulk positively; and AFD and CF, AFD and CRV negatively. The CRV measure is positively associated with crimp frequency but negatively associated with fibre diameter in wool (Fish *et al.*, 1999). The intrinsic resistance to compression of alpaca fibre was low (Lupton *et*

Table 2. Body weight, fleece weight and fibre characteristics by sex (female=632 and male=441).

Breed	Female			Male		
Traits	Mean	SD	n	Mean	SD	n
BWT (kg)	66.6	15.5	308	61.4[**]	17.7	178
FWT (kg)	2.2	0.8	337	2.6[**]	1.0	235
FBW (kg)	1.1	0.4	330	1.4[**]	0.5	230
AFD (μm)	25.8	4.9	621	25.6	4.7	432
FDSD (μm)	5.1	1.2	621	5.1	1.3	432
CV (%)	19.6	2.5	621	20.1[**]	3.2	432
SF (μm)	65.9	35.3	527	64.4	38.8	329
SL (mm)	24.7	4.9	527	24.9	4.7	329
CF (%)	78.9	22.2	527	77.8	21.1	329
FSDL (μm)	1.4	3.7	527	1.3	3.3	329
MFD (μm)	20.1	10.4	527	20.2	10.3	329
XFD (μm)	23.2	10.3	527	23.4	10.6	329
Bulk (cm³/g)	18.3	2.2	526	18.5	2.1	329
HL (mm)	63.6	34.7	522	30.2	36.8	326
CRV (°/mm)	21.5	13.5	527	21.9	12.5	329
CRVD (°/mm)	18.8	8.9	527	19.0	8.5	329
AGE (yr)	4.9	3.7	538	4.2[**]	3.3	336

[**] P< 0.01

BWT: body weight; FWT: fleece weight; AFD: average fibre diameter; FBT: fleece blanket weight; FDSD: fibre diameter standard deviation; CV: coefficient of variation; CF: comfort factor; SF: spinning fineness; SL: staple length; MFD: minimum fibre diameter; XFD: maximum fibre diameter; FSDL: fibre diameter variation along the length; Bulk: New Zealand bulk measurement; HL: Hauteur length; CRV: curvature; CRVD: curvature variation; ns: not significant.

al., 2006), which is in agreement with the current study measured correspondingly as bulk. As a comparison, Huacaya alpacas have fewer staple crimps than the similar micron group of sheep wool, whereas Suri alpaca fibres are straight and nearly free of crimps. Therefore, Suri alpacas tested significantly lower in bulk measurement than Huacayas. All other correlation coefficients were found to be low in value or negligible regardless of direction.

Table 3. Body weight, fleece weight and fibre characteristics by age groups.

Traits	Age group							
	i	ii	iii	iv	v	vi	vii	viii
BWT (kg)	43.4[a]	55.1[b]	65.6[c]	70.8[d]	75.2[e]	78.6[e]	75.4[e]	73.4[de]
FWT (kg)	1.8	2.3	2.6	2.6	2.5	2.6	2.5	2.2
AFD (µm)	22.7	23.5	25.3	26.2	26.5	27.4	28.3	29.1
FBW (kg)	2.2	2.8	2.9	3.0	2.9	3.0	2.9	2.5
FDSD (µm)	4.7	4.8	4.8	5.2	5.2	5.3	5.4	5.9
CV (%)	20.5	20.1	19.3	19.7	19.6	19.3	18.9	20.2
CF (%)	89.3	87.9	81.1	77.9	76.4	73.3	67.3	63.9
SF (µm)	22.0	22.6	24.4	25.3	25.5	26.1	27.2	28.2
SL mm	68.1	71.9	73.8	68.4	70.5	68.5	66.8	64.5
MFD (µm)	16.8	16.4	20.9	19.8	21.3	21.1	25.9	23.4
XFD (µm)	20.5	19.5	23.8	22.8	24.2	24.7	28.3	26.6
SDL (µm)	1.8	1.4	1.1	1.2	1.1	1.8	0.7	1.3
Bulk (cm³/g)	17.8	18.2	18.6	18.3	18.8	18.5	19.2	18.6
HL (mm)	61.8	60.1	67.6	66.2	71.3	67.8	68.8	69.7
CRV (°/mm)	21.4	23.2	23.1	20.5	22.9	20.6	23.4	19.8
CRVD (°/mm)	18.9	20.3	19.5	18.0	19.5	17.9	19.6	17.7

[abcde] Means with a different superscript differs at $P=0.05$ level.
BWT: body weight; FWT: fleece weight; AFD: average fibre diameter; FBT: fleece blanket weight; FDSD: fibre diameter standard deviation; CV: coefficient of variation; CF: comfort factor; SF: spinning fineness; SL: staple length; MFD: minimum fibre diameter; XFD: maximum fibre diameter; FSDL: fibre diameter variation along the length; Bulk: New Zealand bulk measurement; HL: Hauteur length; CRV: curvature; CRVD: curvature variation; ns: not significant.

Conclusions

The body weight, fleece weight and fibre characteristics in two breeds of alpacas farmed in the United States were presented. Fibre characteristics of Huacaya and Suri were found to be similar or slightly higher than their native counterparts in South America, and Suri alpaca fibre was found to be coarser than Huacaya in the U.S. farmed alpacas. Fleece and fibre characteristics of both breeds, Huacaya or Suri are within the desirable characteristics required for textile fibres and home-spun industries.

Table 4. Body weight, fleece weight and fibre characteristic comparison by coat colour groups.

	Coat colour						
	Black	Beige	Brown	Fawn	Grey	L. Fawn	White
BWT (kg)	53.8	66.9	62.4	68.1	49.2	61.9	66.3
FWT (kg)	2.6	2.8	2.2	2.3	2.1	2.5	2.3
AFD (μm)	26.6	25.7	26.6	25.9	27.5	24.8	24.9
FBW (kg)	1.5	1.5	1.1	1.2	1.2	1.3	1.2
FSD (μm)	5.4	5.0	5.5	5.1	5.5	4.9	4.9
CV (%)	20.3	19.4	20.5	19.8	19.9	20.0	19.4
CF (%)	74.2	78.5	75.2	78.4	71.4	83.4	81.1
SF (μm)	25.8	24.7	25.8	25.0	26.6	23.9	24.0
SL (mm)	79.1	64.7	67.5	79.9	79.4	68.3	67.2
MFD (μm)	20.9	17.6	21.2	21.2	24.9	17.8	18.9
XFD (μm)	24.5	21.2	24.5	24.3	28.3	20.9	21.9
FSDL (μm)	1.6	1.9	1.4	1.3	1.5	1.4	1.4
Bulk (cm³/g)	18.6	18.3	18.1	18.5	19.7	18.1	18.5
HL (mm)	77.3	53.9	67.9	68.1	80.0	58.4	63.4
CRV (°/mm)	21.3	21.6	18.8	21.7	25.8	20.8	23.1
CRVD (°/mm)	18.1	18.5	16.6	19.1	21.5	18.9	19.9

BWT: body weight; FWT: fleece weight; AFD: average fibre diameter; FBT: fleece blanket weight; FDSD: fibre diameter standard deviation; CV: coefficient of variation; CF: comfort factor; SF: spinning fineness; SL: staple length; MFD: minimum fibre diameter; XFD: maximum fibre diameter; FSDL: fibre diameter variation along the length; Bulk: New Zealand bulk measurement; HL: Hauteur length; CRV: curvature; CRVD: curvature variation.

Table 5. Correlation coefficients among body weight, fleece weight and selected fibre characteristics.

	FWT	AFD	CV	CF	SL	CRV	Bulk
BWT	0.27^{**}	0.43^{**}	-0.22^{**}	-0.37^{**}	-0.09^{*}	-0.10^{*}	0.02
FWT		0.19^{**}	-0.13^{**}	-0.16^{**}	0.05	0.06	0.14^{**}
AFD			0.00	-0.94^{**}	-0.10^{**}	-0.42^{**}	-0.11^{**}
CV				-0.11^{**}	-0.10^{**}	-0.27^{**}	-0.28^{**}
CF					0.07^{*}	0.34^{**}	0.06
SL						0.33^{**}	0.34^{**}
CRV							0.92^{**}

$^{*}P<0.05$; $^{**}P<0.001$.
BWT: body weight; FWT: fleece weight; AFD: average fibre diameter; CV: coefficient of variation; CF: comfort factor; SL: staple length; CRV: curvature; Bulk: New Zealand bulk measurement.

Acknowledgements

The author would like to thank the alpaca breeders and owners who participated in this study and provided their herd data and fleece samples for the analysis. Huacaya and Suri alpaca production data and fleece samples were collected and provided by the following participants (in alphabetic order): Patti & Alan Anderson (Wild Rose Suri Ranch, MD), Rick & Connie Bodeker (BBF Alpacas INC, MN); Denese & Wick Calahan (Prairie Winds Alpaca Ranch LLC, CO); Nikki & Griffith Collins (Sandollar Alpacas, WA); R.I. Crowe II (Bar C Ranch, NE); Laurie Duff-Robertson (Mountain Silk Alpacas, WA); Eddie & Karil Gray (Grazing Hills Alpacas, ID); Jan & Karl Heinrich (Long Hollow Suri Alpacas, TN); Carrie & Tevis Hull (Timber Basin Ranch Alpacas, ID); Dianne & David Kelley (Kelley Valley Alpacas LLC, ID); Jeannette & David Miller (Sierra Nevada Alpacas, NV); Mary & Stan Miller (Aspen Alpaca Company, ID); Margie Ray (Ray Farms LLC, OK); Beth Roy (Suri Peak Alpacas, CO); Kim & Martin Shelman (Utopia Alpacas, WA); Carol & Rick Thayer (Blonde Velvet & Me, WA); and Brenda & Jeffery Trammell (Enchanted Acres Ranch, WA).

I would like to acknowledge Miss Lei Shi and Miss Shannon McCanahey, my graduate students at the University of Nevada, for their assistance with alpaca fibre measurements. Lincoln University Contribution No. 2010-0011.

References

Cervantes I., M.A. Pérez-Cabal, R. Morante, A. Burgos, C. Salgado, B. Nieto, F. Goyache and J.P. Gutierrez, 2010. Genetic parameters and relationships between fiber and type traits in two breeds of Peruvian alpacas. Small Ruminant Research, 88: 6-11.

Fish E.E., T.J.Mahar and B.J. Crook, 1999. Fiber curvature morphametry and measurements. Wool Technology & Sheep Breeding, 47: 248-265.

Frank E.N., M.V.H. Hick, C.D. Gauna, H.E. Lamas, C. Renieri and M. Antonini, 2006. Phenotypic and genetic description of fiber traits in South American domestic camelids (llamas and alpacas). Small Ruminant Research, 61: 113-129.

Kadwell M., M. Fernandez, H.F. Stanley R. Bald, J.C. Wheeler, R. Roadio and M.W. Bruford, 2001. Genetic analysis reveals the wild ancestress of the llama and the alpacas. Proceedings of the Royal Society of London B268, pp. 2575-2584.

Lupton C.J., A. McColl and R.H. Stobart, 2006. Fiber characteristics of the huacaya alpaca. Small Ruminant Research, 64: 211-224.

McColl A., C. Lupton and R. Stobart, 2004. Fiber characteristics of U.S. huacaya alpacas. Alpaca magazine (summer): 186-196.

McGregor B.A. and K.L. Butler, 2004. Sources of variation in fiber diameter attributes of Australian Alpacas and implications for fleece evaluation and animal selection. Australian Journal of Agriculture Research, 55: 433-442.

Morante R., F. Goyache, A. Burgos, I. Cervantes, M.A. Pérez-Cabal, and J.P. Gutierrez, 2009. Genetic improvement for alpaca fiber production in the Perivian Altiplano: The Pacomarca experience. Animal Genetic Resources Information (FAO of the United Nations) 45: 37-43.

Pumayalla A. and C. Leyva, 1988. Production and technology of the alpaca and vicugna fleece. Proceedings of the first international symposium on specialty fibers, DWI, Aachen, Germany, pp. 234-241.

Sponenberg D.P., 2010. Suri and huacaya alpaca breeding results in North America. Small Ruminant Research, 93: 210-212.

Wuliji T., G.H. Davis, K.G. Dodds, P.R. Turner, R.N. Andrews and G.D. Bruce, 2000. Production performance, repeatability and heritability estimates for live weight, fleece weight and fiber characteristics of alpacas in New Zealand. Small Ruminant Research, 37: 189-201.

Diversity and comparison of wool parameters in 31 different American and European ovine breeds

P.M. Parés Casanova[1], R. Perezgrovas Garza[2] and J. Jordana Vidal[3]
[1]Dept. de Producció Animal, Ciència i Salut Animal, Universitat de Lleida, Av. Alcalde Rovira Roure, 191. 25198 Lleida, Catalunya, Spain; peremiquelp@prodan.udl.cat
[2]Instituto de Estudios Indígenas, Universidad Autónoma de Chiapas (IEI-UNACH), Centro Universitario Campus III, San Cristóbal de Las Casas, 29264 Chiapas, México
[3]Dep. de Ciència Animal i dels Aliments, Facultat de Veterinària, Universitat Autònoma de Barcelona, Edifici V. 08193 Bellaterra, Catalunya, Spain

Abstract

A study was designed with the aim of establishing phenotypic relationships between certain American and European autochthonous sheep breeds based on wool parameters. Data were collected from 1,143 animals from 31 sheep breeds or varieties. Nine quantitative wool characteristics – length and amount of long, short and kemp fibres, mean fibre diameter and fibres with diameter > 30 μ, and percentage after degreasing with alcohol – were investigated. A principal component analysis (PCA) was carried out with the aim of obtaining the variables exhibiting a pattern of jointly contributing to the total variation. To present data structure and relationship among population accessions, a cluster analysis and a principal coordinates analysis of all the data were conducted. For all analyses, the correlation matrix was utilised. Percentage of fibres and fineness appear as the most useful grouping criteria. The resulting dendrogram showed a 'short coarse wool sheep', a 'medium fine wool sheep' (with different percentages of short fibres that allow them to be divided into three subgroups) and a 'long medium wool sheep'. No 'short fine wool sheep' appeared. Although no phylogenetic information is generated, the results seem to establish different, but more concise groups than those traditionally considered: Merino, Entrefino, Churro and Iberian stock.

Keywords: fibre diameter, clustering analysis, phylogeny, population assignment, sheep breeds

Introduction

There is little published information about comparative wool characteristics for some of the breeds investigated. Moreover, there have been conflicting reports in the literature regarding the relationships between some American and European breeds. In this paper, we present an analysis based on breed wool characteristics with a view to obtaining a deeper insight into relationships within and between breeds.

Materials and methods

Sample collection

Data was collected from 1,143 animals belonging to thirty-one autochthonous breeds or varieties from America and Europe. For a description of the breeds, see Esteban (2003). Management for each breed was different, but each provided a good representation of the standard breed. Samples were taken from the mid-lateral part of each ewe before shearing. Sampled animals were randomly chosen from herds. Each sampled animal was more than 18 months old.

Data collection

Each wool's characteristics were determined using standard methods (see Rojas *et al.*, 2005) at the Wool Quality Laboratory at the Institute for Indigenous Studies, Universidad Autónoma de Chiapas (UNACH), in San Cristóbal de Las Casas, Chiapas (Mexico). Nine wool characteristics were investigated for each sample (Parés, 2008): fibre length (cm), percentage of each type of fibre (long-coarse fibres – outer-coat, short-fine fibres – inner-coat, and kemp), mean fibre diameter (μm), percentage of fibres > 30 μm (F30) and yield after scouring with alcohol (%). No colour or crimp parameters were obtained. Data for each wool's characteristics were analysed for their effects on breed.

Data analysis

To present data structure and relationship among population accessions, a principal component analysis (PCA) and cluster analysis were conducted. A PCA was carried out of all the quantitative data with the aim of obtaining an aggregate of variables exhibiting a pattern of jointly contributing to the total variation. The first three eigenvectors were extracted from the correlation matrix derived from the nine measurements collected. Correlation rather than a correlation matrix was used because the former measures standardised variation and is not affected by unequal variances that occur in data matrices with variables of different absolute size (López-Martín *et al.*, 2006). Character loading (i.e. component correlations) was used to determine the contribution of each variable to the first component vector. In order to organise breeds into feasible groups, a constrained method using Ward's method was utilised with data for most discriminative parameters. A principal coordinate analysis (PCoA) was finally generated from the correlation matrix. The scores of breeds for the first two components were projected. Different PAST – 'Paleontological Statistics Software Package for Education and Data Analysis' (Hammer *et al.*, 2001) programmes were utilised to perform all the analyses.

Results and discussion

Descriptive statistics for each breed are shown in Table 1. Inter-breed correlations for the entire population showed considerable variation (0.060 to 0.7129). There is significant variability among different breeds, which indicates a high response to selection.

The plot of sample means on the first two principal components axes is shown in Figure 1. Component 1, which explains 61.2% of the total sample variance, is the short fibre percentage, whereas Component 2, which explains 17.6% of the total sample variance, is the long fibre percentage and fineness. Table 2 shows the variable contributions, in terms of percentage of total variance explained, to the formation of 1-3 axes, obtained by PCA. According to the PCA, the primary source (PC1) of variation for the sample studied was the percentage of short fibres. Fibre diameter and percentage of long fibres are significant variables of PC2. Percentage of kemp fibres is the significant variable for PC3. These are then useful as grouping criteria, as indicated by some authors (Briggs, 1995; Maddever and Cottle, 1999). Scouring seems not to be an important discriminative parameter.

Using the correlation matrix of the four most discriminative parameters (percentage of each kind of fibre, fibre diameter and F30), a dendrogram was obtained utilising Ward's method (Figure 2). As the figure shows, most of the breeds are grouped into three large clusters, with one isolated breed appearing (CHB). The cluster's coefficient correlation was 0.894.

Table 1. Some descriptive wool traits for the sheep breeds studied.

Breeds	Abbr.	n	Long fibres length (cm)	Short fibres length (cm)	Kemp fibres length (cm)	Long fibres (%)	Short fibres (%)	Kemp fibres (%)	Alcoholic yield (%)	Mean diameter fibre (μ)	F30 (%)
Aranesa	ARA	18	4.1	4.1	1.2	5.85	93.17	0.96	84.2	29.5	39.4
Berberina	BER	15	2.5	2.9	0.8	0.75	98.35	0.88	81.1	26.7	23.4
Blanca Colombia	BCO	45	20.8	10.5	4.1	15.23	80.90	3.85	83.3	31.2	40.8
Castellana Negra	CAN	4	6.3	3.6	1.7	19.66	71.84	8.49	75.6	19.2	10.5
Castillonesa	CAS	15	4.3	4.7	0.7	2.49	96.59	0.91	83.5	35.3	66.8
Chiapas blanca	CHB	48	21.9	11.3	6.1	57.87	38.89	3.22	83.4	30.4	40.0
Chiapas café	CHC	51	24.9	12.0	5.5	21.25	74.93	3.81	87.8	29.6	40.0
Chiapas negra	CHN	62	19.7	11.5	8.5	27.11	70.83	2.04	83.9	25.0	27.5
Churra	CHU	25	13.7	7.6	3.2	6.55	85.21	8.23	81.0	24.5	20.5
Churra badana	CBA	74	22.6	15.6	3.5	11.07	77.44	1.48	79.1	39.7	63.6
Churra terra quente	CTQ	30	20.4	11.5	3.4	16.99	76.55	6.44	81.4	35.4	42.7
Criolla de Bolivia	CBO	82	13.5	10.6	2.1	23.12	73.56	3.30	70.3	29.7	38.3
Crioula Brasil	CBR	52	26.8	11.8	4.3	21.56	75.02	3.41	98.7	31.8	38.6
Guirra	GUI	15	7.5	6.3	2.4	3.94	92.36	3.69	74.7	27.7	19.4
Latxa Chilena	LAC	88	24.5	11.3	7.6	17.89	80.47	1.61	89.0	42.9	62.9
Linca	LIN	10	18.6	9.3	0.0	8.34	91.65	0.00	75.5	26.6	19.2
Manchega Negra	MAN	63	9.5	7.6	2.6	0.33	89.92	9.73	77.8	29.0	31.0
Mer. Grazalema	GRA	44	7.7	6.6	1.3	33.28	65.60	1.11	77.3	32.6	46.1
Mirandesa	MIR	40	17.5	12.5	2.9	5.14	83.56	11.28	86.9	30.4	40.5
Mixteca	MIX	61	5.2	3.3	1.2	11.31	87.46	1.21	71.2	23.4	15.0
Mondegueira	MON	28	17.86	8.8	3.2	16.90	75.62	7.46	74.6	38.9	56.5
Mora Colombiana	MOR	27	14.5	8.4	1.9	3.01	96.92	0.05	79.5	20.2	7.98
Navajo-Churro	NAV	29	21.2	11.8	1.8	12.49	86.92	0.57	62.9	24.2	19.5
Oaxaca	OAX	56	9.2	5.5	2.6	19.83	77.34	2.82	75.9	21.5	16.6
Ripollesa	RIP	14	5.5	5.3	1.3	0.30	96.76	2.92	85.5	28.8	29.0
Roja Mallorquina	RMA	12	12.0	9.5	2.9	10.33	83.42	6.24	76.5	27.9	27.4
Socorro	SOC	49	0.0	6.6	0.1	0.71	97.80	1.48	54.7	22.1	1.60
Tarahumara	TAH	44	12.2	8.3	4.3	13.67	82.01	4.30	73.1	26.9	25.3
Tarasconesa	TAR	18	2.5	4.4	1.1	0.00	98.83	1.16	71.4	27.3	32.3
Xisqueta	XIS	10	11.7	4.0	1.4	0.12	98.06	1.80	80.9	30.6	42.3
Zongolica	ZON	14	13.7	9.3	3.0	9.97	87.56	2.46	82.4	24.4	23.9

The results of the PCoA are shown in Figure 3. The results observed in the PCoA analysis reinforce the results observed in the dendrogram. A cluster formed by GRA, LAC, MON, CBA and CAS presents a large fibre diameter (greater than 50 microns) and a medium percentage of long fibres (X=16.3%), allowing it to be labelled 'short coarse wool sheep'. The presence of GRA in this group is not surprising, as it is said that the breed originated as a cross between the Merino and Churra, the latter being the cause of the coarse wool (Esteban, 2003).

It should be noted that CHB differed somewhat from other breeds, with more abundant long fibres (X=57.9%) which are also finer (X=30.4 μ, median=24.0 μ, highly skewed to the right); we

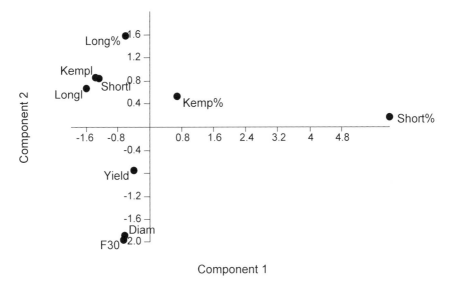

Figure 1. Plot of the sample means for nine wool parameters on the first two principal component axes.

Table 2. Variable contributions, in terms of percentage of total variance explained, to the formation of 1-3 axes, obtained by PCA.

	Axis 1	Axis 2	Axis 3
Long fibres length	-1.615	0.666	-0.129
Short fibres length	-1.297	0.840	0.554
Kemp fibres length	-1.382	0.850	-0.509
Long fibres %	-0.614	1.582[a]	-0.959
Short fibres %	5.996[a]	0.171	-0.442
Kemp fibres %	0.681	0.518	2.300[a]
Alcoholic yielding %	-0.417	-0.769	-0.970
Diameter μ	-0.661	-1.894[a]	0.172
F30%	-0.691	-1.965 [a]	-0.017

[a]Highest contribution of variables (4 out of 9) to axes 1 and 2.

could classify it as a 'long fine wool sheep' with a clear double coat (57.9% long fibres vs. 38.9% short fibres). The 'short coarse wool sheep' and CHB occupy opposite sides of the dendrogram obtained. The middle groups could be classified as 'medium fine wool sheep', with a cluster formed by TAR, RIP, XIS, MAN, MIR, RMA, ARA and BER and another formed by CHC, CBR, BCO, CTQ, CBO and CHN. They are characterised by highly uniform medium fine fibres (X=29.5 μ, median=29.6 μ), but with different percentages of fibres: more abundant long fibres in the latter group (X=20.8%, 'long medium fine' group) than in the former (X=2.8%, 'short medium fine' group). According to Rojas et al. (2005), MIR, BCO, CTQ, NAV, CHU and CHI have similar origins. In the last group (OAX, ZON, TAH, NAV, LIN, MIX, MOR, CHU, GUI, CAN and SOC), there are breeds with values that could allow them to be classified as 'middle medium fine wool sheep' (X=23.7 μ, median=24.3 μ and 9.9% of long fibres). Interestingly, the

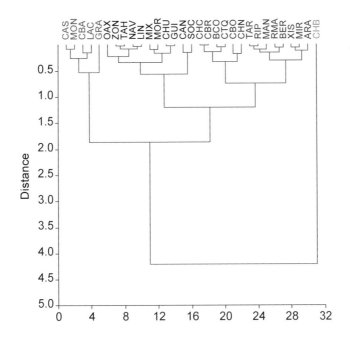

Figure 2. Constrained clustering (Ward's method) (coefficient correlation = 0.894).

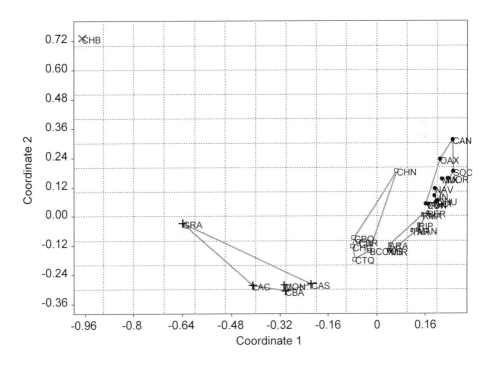

Figure 3. Two-dimensional plot of the principal coordinates from the correlation matrix.

Churra (CHU) and Latxa breeds (LAC), which belong to what is referred to as 'Latxas', were grouped into different clusters. As we have seen, LAC shows a very different pattern of fibres (low fineness and a medium percentage of long fibres). All these 'medium sheep' present a low degree of differentiation in the fleece. Although some of them are still used for traditional textile purposes in Indian villages using traditional textile processes, they do not possess a double coat, as occurs in CHB, according to the bibliography consulted (for instance, see Corzo *et al.*, 2005 for Zongolica). No 'short fine wool sheep' (with a very small fibre diameter, 20 microns or less) appeared.

Conclusion

The data reported here provide valuable insight into population and help in assessing inter-breed comparison, supporting the idea that the analysis of racial relationships is not necessarily linked to historical relationships. However, our results seem to determine different but more concise groups than those traditionally considered: Merino, Entrefino, Churro and Iberian stock.

Acknowledgements

Very special thanks are also extended to the many farmers and owners who made it possible to collect samples from their herds.

References

Briggs P.E., 1995. Evaluating wool on the live animal. The Marker. December: 4-5.

Corzo J., E. Citlahua, D. Galdámez, R. Perezgrovas, S. Vargas, A. Gil and W. Pittroff, 2005. Caracterización de la lana en ovejas autóctonas criadas por indígenas Nahuas de la Sierra de Zongolica, Veracruz (México). In: Proceedings of 'VI Simposium Iberoamericano Sobre la Conservación y Utilización de Recursos Zoogenéticos'. San Cristóbal de las Casas Chiapas. México, pp. 287-289.

Esteban C., 2003. Razas ganaderas españolas. II. Ovinas. FEAGAS. Madrid, Spain.

Hammer Ø., D.A.T. Harper and P.D. Ryan, 2001. Past: paleontological statistics software package for education and data analysis. Palaeontologia Electronica 4 (1), available online: http://palaeo-electronica.org/2001_1/past/issue1_01.html.

López-Martín J.M., J. Ruiz-Olmo and I. Padró, 2006. Comparison of skull measurements and sexual dimorphism between the Minorcan pine marten (*Martes martes minoricencis*) and the Iberian pine marten (*M. m. martes*): a case on insularity. Mammalian Biology, 71: 13-24.

Maddever D. and D. Cottle, 1999. Product-process Groups and wool price. Wool Technology and Sheep Breeding, 47: 38-46.

Parés P.M., 2008. Caracterització estructural i racial de la raça ovina Aranesa. Thesis, Universitat Autònoma de Barcelona, Barcelona, Spain.

Rojas A.L., R. Perezgrovas, G. Rodríguez, P. Russo-Almeida and H. Anzola, 2005. Caracterización macro y microscópica de la lana en ovinos autóctonos iberoamericanos de vellón blanco. Archivos de Zootecnia, 54 (206-207): 477-483.

Testing objective metrics for the differentiation of coat colours in a Spanish alpaca population

E. Bartolomé[1], M.J. Sánchez[1], F. Peña[2] and A. Horcada[1]
[1]*Dpto. Ciencias Agroforestales – Área de Producción Animal. EUITA. Universidad de Sevilla. Ctra. Utrera, km 1 – 41013, Sevilla, Spain; ebartolome@us.es*
[2]*Dpto. Producción Animal. Facultad de Veterinaria. Universidad de Córdoba. Campus de Rabanales. Ctra. Madrid-Córdoba, km 396ª, Córdoba, Spain*

Abstract

Objective measurement of alpaca coat colour is now an integral part of the process of showing, breeding, marketing and classing of alpacas. The aim of this study was to test the capacity of the CIELab Colourimetric System to differentiate between coat colour classes defined by official patterns in alpacas. These classes were defined regarding to different pre-established subjective patterns (British and Australian official patterns, the ICAR pattern proposal) and the colour estimated by the researcher in the field), and compared with quantitative colourimetric measurements according to a standardised international procedure. A total of 231 alpacas from Suri (24%) and Huacaya (76%) breeds, were measured at shoulder, ribs and croup for colourimetric parameters and classified according to these four colour patterns. The discriminant analysis showed that, while black and white colours were the best assigned coats by observation (at 94.7% and 92.2%, respectively), the colourimeter assigned more coats to these patterns than those identified by observation (30 and 74 assigned by colourimeter, over 19 and 64 identified, for black and white coat colours, respectively).

Keywords: CIELab Colourimetric System, *Lama pacos*, colour chart, fleece classification

Introduction

Alpacas (*Lama pacos* L.) are domestic South American camelids, located mainly in the Andean regions of Bolivia and bred for fibre and meat production (Bonavia, 1996). Although camelids were regarded as being exclusively adapted to this environment, introduction to various other countries has been highly successful (Wuliji *et al.*, 2000) enhancing this market all over the world. However, the main interest of these countries in these animals is as fibre producers and pets.

Alpaca fibre is especially appreciated for the large number of natural colours, ranging from white to black with several intermediate levels (FAO, 2005; Frank *et al.*, 2006). Several patterns have been developed to classify alpaca fleece in terms of fibre colour, following either subjective (regarding to visual parameters of a trained judge that ascribes the fibre colour following an official colour pattern panel (Bustinza, 2001; Schmid, 2006)) or objective methods of classification (according to 'Munsell Colour System' (Ruiz de Castilla and Mamani, 1990; Renieri *et al.*, 1991), using near-infrared reflectance spectrometer or SAMBA methods (Lupton and McColl, 2008) or according to pigment concentration measurement (Toth *et al.*, 2006)). Therefore, an objective measurement method that could accurately ascribe fleece coat colour according to official colour panels without the assessment of a trained judge, would improve the reliability of the measures.

The Commission Internationale de l'Eclairage L*a*b* system is a colourimetric measurement method that was specified for the representation of a uniform colour space that could approximate human vision. It was based on the 'Munsell Colour System' accounting for a wider colour spectrum. Furthermore, the CIE colour method has previously been used to test fibre colour in alpacas, proving to be quite accurate in differentiating main colour groups (Bartolomé *et al.*, 2009). In this regard, the aim of this study was to test the discrimination capacity of the CIE L*a*b* Colourimetric System to differentiate between coat colour classes defined by official patterns in alpacas.

Materials and methods

A total of 231 alpacas from 5 different farms of the Spanish population were measured, 55 from the Suri breed and 176 from the Huacaya. Coat colour of every animal was categorised according to five different pre-established subjective patterns: British (12 classes), Australian (12 classes) and Peruvian (13 classes), the official ICAR pattern proposal (Antonini, 2006) that described 15 classes and the colour estimated by the researcher in the field, with 6 classes (Table 1).

Coat colour was also measured quantitatively according to a standardised international method: the Commission Internationale de l'Eclairage L*, a*, b* (CIELab), where L* describes lightness, with values from 0 (black) to 100 (white); a* describes colour saturation from red to green on a scale of +60 to -60, where positive values indicate varying intensities of red; and b* describes colour saturation from yellow to blue on a scale of +60 to -60, where positive values indicate varying intensities of yellow. For the colourimetric parameters, each animal was measured alive on three different body regions: shoulder, ribs and croup, taking the colourimeter measurement from the base of the fibre. For the colourimetric measurements a Minolta ChromaMeter model CR-400-410 was used.

A univariate (ANOVA) and a multivariate (MANOVA) variance analysis were performed to attempt to find significant differences between coat colours according to colourimetric parameters.

Table 1. Description of the different colour charts used to define the coat colour of the Spanish alpaca population.

British pattern	Australian pattern	Peruvian pattern	ICAR pattern	Researcher
white	white	white	white	white
light fawn	light fawn	light cream	light fawn	fawn
medium fawn	medium fawn	medium cream	light brown	brown
dark fawn	dark fawn	dark cream	self brown	black
light brown	light brown	brown[l]	dark brown	mixed
medium brown	medium brown	light coffee	black	grey
dark brown	dark brown	medium coffee	light grey[l]	
black	black	dark coffee	self grey[l]	
light grey[l]	light grey[l]	black	dark grey[l]	
medium grey[l]	medium grey[l]	light grey[l]	light roan[l]	
dark grey[l]	dark grey[l]	undefined grey	self roan[l]	
rose grey/roan	rose grey/roan	dark grey[l]	dark roan[l]	
		roan[l]	light pink[l]	
			self pink	
			dark pink[l]	

[l] Colour levels not included in the analysis due to not being present in the population.

Accuracy of colour assignation according to the pattern chart used and to the CIELab colour system, was assessed by a classification matrix from a discriminant analysis. All the statistical analyses were performed using the Statistica v6.0 software.

Results and discussion

Although direct visual appraisal of coat colour pattern, carried out with other objective description methods such as the Munsell Colour System, are not very accurate and should be supported by other objective methods (Cecchi *et al.*, 2001; Lauvergne *et al.*, 1996), the MANOVA and ANOVA analyses showed significant statistical differences for all the colour levels described in every colour chart assessed, according to the 3 colourimetric parameters (L*, a* and b*) recorded per animal. This could be due to a more complete colour space specified by this method than the Munsell Colour system.

Regarding the discriminant ability of the CIELab Colour System to classify fibre colours according to 5 different pre-established colour charts, the accuracy highlighted was lower than would be expected from the previous results (Table 2), with the 'Researcher pattern' the chart that showed the highest percentage of correct assignations with a 68.8% score, followed by ICAR pattern with 57.9%, the British and Peruvian pattern with 50% each and finally, the Australian pattern with 45.7%. Despite that the percentage of correct assignations showed medium to moderate high values, the reliability of the system appeared to be quite low because for every pool of coats assigned, neither of the patterns studied could assure less than 25% false positives, considerably decreasing the accuracy of the test (Falconer and Mackay, 1996). However, these results highlighted the fact that 'Researcher' and 'ICAR' patterns were the colour classifications

Table 2. Classification matrix (discriminant analysis) for five coat colour patterns, analysed according to colourimeter parameters (L, a* and b*).*

Colours	Percentage of correct assignations	Assigned by the colour pattern	Assigned by the colourimeter
Researcher pattern			
White	92.2	64	74
Fawn	55.9	34	28
Brown	80.3	66	85
Black	25	4	1
Grey	94.7	19	30
Mixed	20.5	44	13
Total	68.8	-	-
British pattern			
White	62	78	90
Light fawn	25	48	22
Medium Fawn	61.5	39	62
Dark fawn	17.7	17	11
Light brown	0	14	0
Medium Brown	71.4	27	43
Dark brown	96.3	27	43
Black	0	9	0
Rose grey	29.4	12	7
Total	50	-	-

Table 2. Continued.

Colours	Percentage of correct assignations	Assigned by the colour pattern	Assigned by the colourimeter
Australian pattern			
White	66.3	80	100
Light fawn	6.7	45	10
Medium Fawn	68.9	45	75
Dark Fawn	0	15	5
Light Brown	6.3	16	5
Medium Brown	57.6	33	44
Dark brown	89.5	19	35
Black	0	8	1
Rose grey	17.7	17	3
Total	45.7	-	-
Peruvian pattern			
White	63.8	80	96
Light cream	26	50	26
Medium Cream	51.4	37	51
Dark cream	8.3	12	4
Light coffee	7.1	14	2
Medium coffee	84.4	32	54
Dark coffee	76	25	32
Black	45.5	11	10
Undefined grey	17.7	14	3
Total	50	-	-
ICAR pattern			
White	58.2	79	78
Light fawn	64.6	82	82
Light brown	38.2	34	34
Self brown	66.7	30	40
Dark brown	92.9	28	41
Black	0	8	0
Self pink	17.7	17	3
Total	57.9	-	-

that could be best described by colourimeter parameters. This could be due to the fact that these charts showed the least number of colour classes and thus would be easy to fit with intervals from the continuous distribution represented by the CIELab.

Regarding the 'Researcher pattern', beside the fact that black and white colours were the best assigned coats by observation (in a 94.7% and 92.2%, respectively), the colourimeter assigned more coats to these patterns than those identified by observation (30 and 74 assigned by colourimeter, over 19 and 64 identified, for black and white coat colours, respectively). On the other 4 colour charts, brown colours appeared to be the best assigned coats by observation, with the colourimeter as the method that assigned more coats to these patterns. Thus, for the 'British pattern', medium brown and dark brown colours showed 71.4% and 96.3% of correct assignations (respectively) and 16 more coats assigned by colourimeter than by visual observation (in both cases); for the 'Australian pattern', dark brown colour showed 89.5% of correct assignations and also 16 more coats assigned by colourimeter; for the 'Peruvian pattern', medium and dark coffee colours showed the highest percentages (84.4% and 76%, respectively), with 22 and 7 more coats assigned by colourimeter (respectively); and finally, dark brown was the coat that showed the highest percentage of correct assignations (92.9%) from the ICAR Pattern, with 13 more coats assigned by colourimeter. This last pattern also showed the same number of coats assigned either by the colour pattern and the colourimeter, for light fawn (82) and light brown (34) colours.

Despite the fact that 'Researcher pattern' seems to be the method that could be better described by colourimeter parameters, a more accurate method would be more desirable because most of the pattern charts used at the present time for fibre colour classification attempts for more colour classes (Wuliji *et al.*, 2000; FAO, 2005). In this regard, the 'ICAR pattern' that ascribe more colour classes, appeared also to be well described by the CIELab colourimetric system. In 2007 the International Committee for Animal Recording (ICAR) set up a Working Group on Animals for Fibre, with the main objective of providing a common approach to Alpaca breed management paying particular attention to fibre production (Antonini, 2006). Thus, the results of this study highlighted the colour pattern proposed by this organisation as a desirable classification to be used by the CIELab colour system, not forgetting the limits assessed here, as only 58% of the coat colours would be correctly assigned.

Conclusion

The results highlighted that the CIELab colour system seems to be a useful method for differentiating coat colours in alpaca fibre. Otherwise, the reliability of the results appeared not to be very high, as the percentage of false positives (coats wrongly assigned to one colour) would be high. Regarding the official patterns studied, the 'Researcher pattern' seems to be most accurate, despite the fact that it showed few colour classes. On the other hand, this study has highlighted the ICAR pattern proposal as a moderately useful pattern to classify alpaca fibre colours with the CIELab colour system.

Further studies should be carried out in order to find an objective measurement method that could account for all the coat colour classes currently described by the market fleece colour patterns just as differences in coat colours according to the alpaca breed (Suri or Huacaya).

Acknowledgements

The authors wish to thank the Spanish Alpaca breeders for their collaboration in providing the information for this study.

References

Antonini M., 2006. ICAR guidelines for alpaca shearing management, fibre harvesting and grading. In: proceedings of 36th ICAR Session. Niagara Falls, Ontario, Canada, 16-20 June 2006.
Bartolomé E., F. Peña, M.J. Sánchez, J. Daza, A. Molina and J.P. Gutiérrez, 2009. Genetic analysis of colorimetric parameters for the differentiation of coat colours in the Spanish Alpacas population. 60th Annual Meeting of the European Association for Animal Production. August 23rd – 27th, Barcelona, Spain.
Bonavia D., 1996. Los Camelidos Sudamericanos (South American Camelids). IFEA-UPCH-Conservation International, Lima, Peru, 843 pp.
Bustinza V., 2001. La alpaca, conocimiento del gran potencial andino (The alpaca, knowledge of the great andine potential). Libro 1. Oficina de Recursos de Aprendizaje, Universidad Nacional del Altiplano, Puno, Peru, 496 pp.

Cecchi T., P. Passamonti, E.N. Frank, M. Gonzales, F. Pucciarelli and C. Renieri, 2001. Pigmentation in South American camelids. Part I. Quantification and variation of combined eumelanins and pheomelanins in various coat colours. In: M. Gerken and C. Renieri (eds.), Progress in S.A. Camelids Res. EAAP. Pub. No. 105, pp. 207-210.

Falconer D.S. and F.C. Mackay, 1996. Introduction to Quantitative Genetics. Longman Group Ltd, Essex, UK. 469 pp.

FAO (Food and Agriculture Organization), 2005. Situación Actual de los Camélidos Sudamericanos en el Perú. (Present situation of South American camelids in Peru). Proyecto de Cooperación Técnica en apoyo a la crianza y aprovechamiento de los Camélidos Sudamericanos en la Región Andina TCP/RLA/2914. Available at: www.rlc. fao.org/es/ganaderia/pdf/2914per.pdf.

Frank E., M. Hick, C. Gauna, H. Lamas, C. Renieri and M. Antonini, 2006. Phenotypic and genetic description of fibre traits in South American domestic calmelids (llamas and alpacas). Small Ruminant Research, 61: 113-129.

Lauvergne J.J., C. Renieri and E.N. Frank, 1996. Identification of some allelic series in Domestic Camelids of Argentina. In: M. Gerken and C. Renieri (eds.), 2nd European Symposium on South American Camelids, pp: 39-50.

Lupton C. and A. McColl, 2008. Evaluation of two objective methods for measuring luster in Suri alpaca fibre, comparison with subjective luster assessments, and correlation with other physical properties. Suri Network News, April 2009, pp 13-14.

Renieri C., M. Trabalza Marinucci, G. Martino and G. Giordano, 1991. Preliminary report on qualità of hair and coat colour in pigmented alpaca. Proceedings IX Italian ASPA, pp. 905-914.

Ruiz de Castilla M. and N. Mamani, 1990. Estudio preliminar del color de la fibra de Llama en los distritos de Callalli y Tisco – provincia de Caylloma_ Arequipa (Preliminary studies of Lama fibre colour in Callalli and Tisco districts). In: Informe de trabajos de investigación en Alpacas y Llamas de color. Volume I (fibras), pp. 1-18.

Schmid S., 2006. The value chain of alpaca fibre in Peru, an economic analysis. Master Thesis, Institut für Agrarwirtschaft, ETH Zürich, Switzerland.

Toth Z., M. Kaps, J. Sölkner, I. Bodo and I. Curik, 2006. Quantitative genetic aspects of coat colour in horses. Journal of Animal Science, 84: 2623-2628.

Wuliji T., G.H. Davis, K.G. Dodds, P.R. Turner, R.N. Andrews and G.D. Bruce, 2000. Production performance, repeatability and heritability estimates for live weight fleece weight and fibre characteristics of alpacas in New Zealand. Small Ruminant Research, 37: 189-201.

Breeding and genetics

Analysis of the mitochondrial diversity of alpacas in eight farming areas of the south of Peru

C. Melo[1], A. Manunza[1], M. Melo[2], L. Olivera[2] and M. Amills[1]
[1]Departament de Ciència Animal i dels Aliments, Universitat Autònoma de Barcelona, 08193 Bellaterra, Spain; Carola.Melo@uab.cat
[2]Facultad de Medicina Veterinaria y Zootecnia, Universidad Nacional del Altiplano, 0051 Puno, Peru

Abstract

Alpaca (*Lama pacos*) is a domesticated South American camelid mainly distributed in Peru, although also present in Bolivia, Chile and Ecuador. This species, highly appreciated because of its lustrous and silky natural fibre, has been poorly studied at the genetic level. Characterisation of the genetic diversity of alpaca would be fundamental for gaining new insights into past domestication events as well as for understanding the demographic history of this species. In the present work, we report the mitochondrial genetic variability of alpacas (n=29) distributed in eight farming areas of the south region of Peru. Sequencing of the D-loop region revealed the existence of 16 haplotypes. Haplotype (Hd) and nucleotide (π) diversities reached values of 0.96 and 0.0456, respectively. Moreover, analysis of the cytochrome b region evidenced the segregation of 15 haplotypes (Hd = 0.946, π = 0.0279). We can conclude that, despite the drastic reduction in the census of alpacas, there is still a considerable level of genetic variation. Median-joining network and neighbour joining phylogenetic analyses showed the existence of two guanaco-like and vicuña-like mitochondrial haplogroups. These results might be explained by the occurrence of ancient hybridisation events between llamas and alpacas with the goal of recovering the alpaca population that was decimated by the Spanish colonisers two centuries ago.

Keywords: alpaca, D-loop, cytochrome b, diversity

Introduction

Alpaca is one of the most important domestic animals in the Andean range (Peru, Chile and Bolivia), occupying mountainous areas at an altitude of 3,800-5,000 meters and feeding on natural grass. It is estimated that the worldwide census of alpacas is 3.6 million, being mostly distributed in Peru (80% of this census) and, within this country, at Puno (www.inia.gob.pe). Alpacas were domesticated approximately 6,000 YBP by the Incas. The fine quality of its fleece, its sociable character and intelligence and its easy handling makes the alpaca not only a valuable domestic animal but also a popular pet. As of 1532 (Spanish colonisation), and with the introduction of Iberian domestic animals, the native flocks of alpaca were decimated. In this way, at least 90% of alpacas and llamas disappeared one hundred years after the arrival of the Spanish colonisers (Flores Ochoa, 1982), being displaced to great altitudes where foreign domestic animals could not be raised (Novoa and Wheeler, 1984; Sumar and García, 1987). Another consequence of the Spanish colonisation was that the admixture of wild and domestic camelids became more and more frequent, and genetic patterns of each species were progressively 'diluted'. In the case of alpaca, this involved a decrease in fibre quality (Kadwell *et al.*, 2001). In the last years, efforts have been made to preserve the genetic reservoir of alpacas. One of the main lines of action of conservation programmes is to characterise the genetic diversity of this species with molecular markers. Herewith, we have analysed the variability of the mitochondrial DNA of Huacaya and Suri alpacas distributed in eight farming areas of Southern Peru.

Materials and methods

Genomic DNA was extracted from 29 samples of hair taken from alpacas representative of the Huacaya (27) and Suri (2) types. Sampled regions were chosen according to their census, phenotypic and production specificities and other criteria (Figure 1, Table 1).

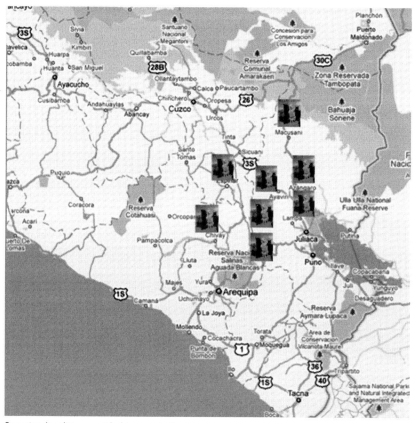

Figure 1. Peruvian localities sampled in our study.

Table 1. Geographical distribution of the alpacas sampled in the current study.

Sampled regions	Sample size CytB	Sample size D-loop
Nuñoa-Puno	-	2
Azángaro-Puno	3	4
Lampa-Puno	5	4
Santa Lucía-Puno	6	4
Ocuviri-Puno	3	3
Macusani-Puno	3	3
Caylloma-Arequipa	7	4
Layo-Cusco	2	3
Total	29	27

Fragments corresponding to the totality of the control (D-loop, 1484 bp) and Cytochrome b (CytB, 1355 bp) mitochondrial regions were amplified by PCR. Primers were designed based on GenBank sequence NC_002504. PCR products were purified with the kit Exo-SAP-IT kit (GE Healthcare) and sequenced with the BigDye Terminator v.3.1 Cycle Sequencing kit (Applied Biosystems). Sequencing reactions were denatured and electrophoresed in an ABI PRISM 3730 device (Applied Biosystems). Mitochondrial sequences were analysed with the Sequencing Analysis v.5.1.1 (Applied Biosystems) software and the alignment of nucleotide sequences for the two genes was carried out using Muscle, implemented in Bosque v. 1.7.157 (Ramírez-Flandes and Ulloa, 2008). The DnaSP v.4.5 programme (Rozas *et al.*, 2003) was used to calculate nucleotide (π) and haplotype (Hd) diversities. The Network v.4.5.1.0. software (Fluxus Technology Ltd) was employed to construct a median joining network including all mitochondrial sequences. Bosque v. 1.7.157 was used to find out the best model for our dataset, which is HKY+G as defined with the JModeltest 0.1 application. Data were clustered using the Maximum Likelihood statistical method and then bootstrapped as a measure of robustness. In addition, the resulting tree was edited using MEGA 4.1 (Tamura *et al.*, 2007).

Results and discussion

Sequencing of the alpaca D-loop region (D-loop) showed the existence of 16 haplotypes with high haplotypic (Hd = 0.96) and nucleotide (π = 0.0256) diversities. Similarly, cytochrome B analyses evidenced the existence of 15 haplotypes with high haplotypic (Hd= 0.946) and nucleotide (π = 0.0279). These results are similar to previous data obtained in other camelid species with microsatellite markers (Penedo *et al.*, 1998; Sasse *et al.*, 2000, Sarno *et al.*, 2001; Wheeler *et al.*, 2001; Maté *et al.*, 2004 and Marín *et al.*, 2007). We can conclude that, in spite of the bottlenecks and founder effects that this species has undergone in the last centuries, there is still a considerable level of genetic diversity. As shown in Figure 2, maximum likelihood phylogenetic analysis of CytB sequences has allowed us to identify two main mitochondrial lineages corresponding to vicuña and guanaco. Similar results were obtained when performing median joining network analyses (Figure 3). Phylogenetic analyses of D-loop sequences were also fully consistent with CytB data (results not shown).

The existence of these two main lineages was firstly reported by Kadwell *et al.* (2001) and Wheeler *et al.* (2001). They explained this finding on the basis of an introgression of llamas (that harbour 'guanaco-type' mitochondrial haplotypes) within the alpaca gene pool (that it is expected to descend from domesticated vicuñas). If so, the maternal introgression of alpacas with llamas would have been very extensive (in our study, 90% of the CytB sequences are grouped in the guanaco cluster). We can also conclude that there is a substantial gene flow between the seven farming areas analysed in our study (mitochondrial sequences do not cluster according to their geographical location).

In the last two decades, male alpacas have been systematically crossed with female llamas as a way of maximising total production of the highly priced alpaca fleece. Since prices have been mostly dictated by fleece weight, without taking into account fineness, the quality of alpaca fibre has decreased steadily. This farming practice might be the main reason for the high llama introgression of alpacas (Kadwell *et al.*, 2001).

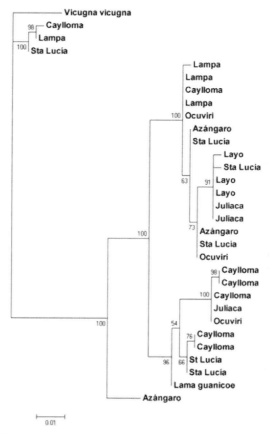

Figure 2. Maximum likelihood phylogenetic tree depicting genetic relationships between alpaca, guanaco and vicuña sequences. Branch lengths are proportional to phylogenetic distances amongst individuals.

Conclusions

The results of this work show that the current population of south Peruvian alpacas displays a high genetic variability, and that alpacas have been strongly introgressed by llamas at least at the maternal level. Although this hybridisation process might have substantially increased alpaca genetic diversity, frequencies of alleles favouring a higher fleece quality might have experienced a drastic reduction as a result of admixture. The identification of alpacas harbouring vicuña-like alleles might be fundamental to implementing strategies aimed at recovering part of the primitive alpaca gene pool.

Acknowledgements

Many thanks to all the Peruvian alpaca farmers who contributed samples to our study.

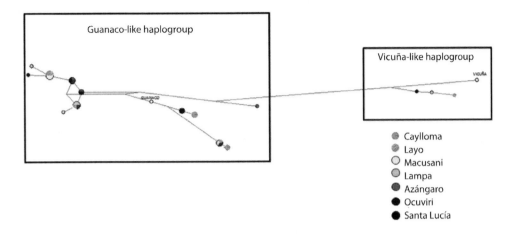

Figure 3. Median-joining network of CytB sequences from alpacas distributed in seven farming areas. Each circle represents a given haplotype, and its size is proportional to haplotype frequency.

References

Flores Ochoa J.A., 1982. Causas que Originaron la Actual Distribución Espacial delas Alpacas y Llamas. In: L. Milliones and H. Tomoeda (eds.), El Hombre y su Ambiente en los Andes Centrales. Osaka: National Museum of Ethnology, Osaka, Senri Ethnological Studies, 10, pp. 63-92.

Kadwell M., M. Fernández, H. Stanley, R. Baldi, J. Wheeler, R. Rosadio and M. Bruford, 2001. Genetic analysis reveals the wild ancestors of the llama and the alpaca. Proceedings of Royal Society of London, 268: 2575-2584.

Penedo M., A. Caetano and K. Cordova, 1998. Microsatellite markers for South American camelids. Animal Genetics, 29: 411-412.

Maté M., F. Di Rocco, A. Zambelli and L. Vidal-Rioja, 2004. Mitochondrial DNA Structure and Organization of the Control Region of South American Camelids. Molecular Ecology Notes, 4: 765-767.

Marín J., C. Casey, M. Kadwell, K. Yaya, D. Hoces, J. Olazabal, R. Rosadio, J. Rodriguez, A. Spotorno, M. Bruford and J. Wheeler, 2007. Mitochondrial phylogeography and demographic history of the Vicuña: implications for conservation. Heredity, 99: 70-80.

Novoa C. and J. Wheeler, 1984. Llama y alpaca. In: I.L. Mason (ed.), Evolution of domesticated animals. Longman, London, United Kingdom, pp. 116-128.

Ramírez-Flandes S. and O. Ulloa, 2008. Bosque: Integrated phylogenetic analysis software. Bioinformatics, 24(21): 2539-2541.

Rozas J., J. Sánchez-Delbarrio, X. Messeguer and R. Rozas, 2003. Bioinformatics, 19: 2496-2497.

Sasse J., M. Mariasegaran, R. Babu, J. Kinne and U. Wernery, 2000. South American camelid microsatellite amplification in *Camelus dromedaries*. Animal Genetics, 31: 75-76.

Tamura K., J. Dudley, M. Nei and S. Kumar, 2007. MEGA4: Molecular Evolutionary Genetics Analysis (MEGA) software version 4.0. Molecular Biology and Evolution, 24: 1596- 1599.

Sarno R., W. Franklin, J. O´Brien and W. Johnson, 2001. Patterns of mtDNA and microsatellite variation in an island and mainland population of guanacos in southern Chile. Animal Conservation, 4: 93-101.

Sumar J. and M. Garcia, 1987. Principios de la Reproducción de la alpaca. IVITA-UNMSM. Lima- Perú, 1: 6-8.

Wheeler J., M. Fernández, R. Rosadio, D. Hoces, M. Kadwel and M. Bruford, 2001. Diversidad genética y manejo de poblaciones de vicuñas en el Perú. Revista Virtual Visión Veterinaria (Perú) 1: 170-183.

Wheeler J., 2006. Informe final, INCAGRO. Proyecto Identificación y rescate de alpacas genéticamente puras de la amenaza de extinción. Lima, Peru.

Asip and *MC1R* cDNA polymorphism in alpaca

C. Bathrachalam, V. La Manna, C. Renieri and A. La Terza
School of Environmental and Natural Sciences, University of Camerino, Via Gentile III da Varano, Camerino (MC), 62032, Italy; chandramohan.bathrchalam@unicam.it

Abstract

Agouti signalling protein (*Asip*) and *Melanocortin 1 Receptor* (*MC1R*) represent key players in coat colour determination in mammals. To investigate the role of these genes in coat colour variation in alpaca, mRNA purified from skin biopsies of different coloured animals were reverse-transcribed into single-stranded cDNAs and the coding sequences (CDS) amplified by PCR, cloned and sequenced. Moreover, and in addition to CDS, the 5'and 3'untranslated regions (UTRs) of both genes were also characterised by RACE experiments. In fact, these regulatory regions can potentially host polymorphisms able to significantly alter *Asip* and *MC1R* gene expression levels. In this study, we described for the first time seven single point mutations in the CDS of *Asip*, of which three were found to be silent mutations and four were missense (amino acid changing) mutations. *MC1R* analysis unveiled a total of ten mutations in the CDS and among those one was 4bp frameshift mutation, four were silent mutations and five were missense mutations. Further analysis of the *Asip* 3'UTR (223 nt) showed three mutations, including one transversion and two transition mutations located at 10, 38 and 77 nt downstream from the stop codon, respectively. Similarly, three transition mutation were also identified in the 3'UTR (602 nt) of *MC1R* located 5,166 and 398 nt downstream from the stop codon. In contrast, two different 5'UTRs were characterised for *MC1R* (294 and 153 nt). Genotyping assays and semi-quantitative RT-PCR analysis of *Asip* and *MC1R* transcripts to evaluate expression-dependent coat colour variation are currently in progress.

Keywords: *Asip*, *MC1R*, cDNA polymorphism, alpaca

Introduction

Pigmentation in mammals is known to be influenced by more than 350 genes (Montoliu *et al.*, 2010). Among these, we find two major genes, namely *Asip* and *MC1R*, which act by regulating the type, amount and distribution pattern of the pigments eumelanin and pheomelanin. Molecular genetics and pharmacological studies have shown that mutually exclusive binding (Ollmann *et al.*, 1998) of MC1R by the agouti protein or by α-melanocyte stimulating hormone (α-MSH) signals hair-bulb melanocytes to synthesise either pheomelanin or eumelanin, respectively. In alpaca, a wide variety of colours exists. Phenotypes, genetics of fibre traits (Frank *et al.*, 2006) and quantitative variation of melanin in alpaca and llamas have been studied in the context of chemical properties of melanins and morphology of melanosomes (Cecchi *et al.*, 2007). Only very limited information is available about the molecular basis of coat colour in alpaca. Therefore, further knowledge of the molecular mechanism behind coat colour variation is needed to assist the breeder in fibre production. In this paper, we investigated and reported for the first time *Asip* and *MC1R* cDNA polymorphism in Peruvian alpaca population.

Material and methods

Collection and storage of skin biopsies

Skin biopsies from coloured (black, white and brown) alpacas were collected in RNAlater (SIGMA, Germany) from Quimsachata Experimental Station, Peru. The biopsies were stored at -196 °C for further analysis.

Primer designing

Using NCBI GenBank, orthologous sequences of *Asip* and *MC1R* from other mammals were collected, aligned with EMBL ClustalW to identify the conserved region which were used to design degenerate primers to amplify the entire *Asip* and *MC1R* coding region.

RNA isolation, cDNA synthesis, cloning and sequencing

The skin biopsies were chopped into small pieces and then subjected to RNA extraction using the RNAeasy fibrous tissue mini kit (Qiagen S.A., Courtaboeuf, France), according to the manufacturer's instructions. cDNA was synthesised with Reverse transcriptase enzyme (TAKARA BIO INC, Japan) followed by RT-PCR. 5' & 3'RACE experiments were carried out according to the method reported in Xianzong and Donald (2006) using Expand long-range dNTPack (Roche, Germany). Then the desired amplicons were gel purified using NucleoSpin Extract II (Macherey-Nagel, Germany), cloned into pGEM-T vector (Promega, USA) and sequenced (BMR genomics, Padova, Italy and StarSeq, Germany).

Results and discussion

At present, we have characterised the whole coding region (CDS) of *Asip* and *MC1R* cDNA in 35 multi-coloured Peruvian alpacas. The complete CDS of *Asip* comprises 402 bp, which is 89%, 88% and 85% identical to cow, sheep and horse respectively. It encodes a putative protein of 133 amino acid (aa). The complete CDS of *MC1R* comprises 954 bp, which is 97%, 90% and 88% identical to camel, dolphin and pig respectively. It encodes a putative protein of 317 aa.

Here, we described for the first time seven single point mutations in the CDS of alpaca *Asip* cDNA, of which three were found to be silent mutations and four were missense mutations (Table 1). *MC1R* analysis unveiled a total of ten mutations in the CDS, among which one was 4bp frameshift mutation, four were silent mutations and five were missense mutations (Table 2). Within the observed *MC1R* mutation, eight of them were already described by Powell *et al*. (2008) and Feely and Munyard (2009) in American and Australian alpaca populations, whereas the remaining two mutations represent novel polymorphisms observed for the first time within the Peruvian population. The analysis of the *Asip* 3'UTR (223 nt) showed three mutations. Among these one was a transversion and two were transition mutations (Table 1) located at 10, 38 and 77 nt downstream from the stop codon, respectively. Similarly, three transition mutations (Table 2) were also identified in the 3'UTR (602 nt) of *MC1R* and were located at 5, 166 and 398 nt downstream from the stop codon. Furthermore, two different 5'UTRs were characterised for *MC1R* (294 and 153 nt). Each one of the identified missense mutations may be involved in coat colour determination.

Table 1. SNPs located in Asip cDNA (CDS and 3'UTR).

Base position	SNPs	Amino acid change	Effect on protein
11	C/G	T/S	Polar
18	A/C	--	N.S.
102	G/A	--	N.S.
152	C/A	Y/S	Polar
290	C/A	--	N.S.
291	T/C	C/T	Slightly polar to polar
352	G/A	H/R	Polar
10 no downstream to the stop codon	C/A	--	--
38 nt downstream to the stop codon	A/G	--	--
77 nt downstream to the stop codon	T/C	--	--

N.S.= non significant.

Table 2. SNPs located in MC1R cDNA (CDS and 3'UTR).

Base position	SNPs	Amino acid change	Effect on protein
82	A/G	T/A	Polar to nonpolar
92[a]	C/T	T/M	Polar to nonpolar
126	C/T	--	N.S.
224-227	ACTT	Frame shift	Frame shift
259[a]	A/G	M/V	Nonpolar to polar
354	T/C	--	N.S.
376	A/G	S/G	Polar to nonpolar
618	G/A	--	N.S.
901	C/T	R/C	Polar to slightly polar
933	G/A	--	N.S.
5 nt downstream to the stop codon	T/C	--	--
166 nt downstream to the stop codon	C/T	--	--
398 nt downstream to the stop codon	G/A	--	--

N.S.=non-significant.
[a] Novel SNP observed in the Peruvian population

Conclusion

These findings will be helpful in understanding the regulatory mechanism of *Asip* and *MC1R* gene in alpaca coat colour variation. Genotyping assays and semi-quantitative RT-PCR analysis of *Asip* and *MC1R* transcripts to evaluate the association between alleles and coat colour and expression-dependent coat colour variation are currently in progress.

Acknowledgements

We would like to thank the Alpaca Research Foundation (ARF) for their financial support of this research.

References

Cecchi T., A. Valbonesi, P. Passamonti, E. Frank and C. Renieri, 2007. Quantitative variation of melanins in llama (*Lama glama* L.), Small Ruminant Research, 71: 52-58.

Frank E.N., M.V.H. Hick, C.D.Gauna, H.E. Lamas, C. Renieri and M. Antonini, 2006. Phenotypic and genetic description of fibre traits in South American domestic camelids (llamas and alpacas). Small Ruminant Research, 61: 113-129.

Feely N.L. and K.A. Munyard, 2009. Characterisation of the melanocortin-1 receptor gene in alpaca and identification of possible markers associated with phenotypic variations in colour. *Animal Production Science,* 49: 675-681.

Montoliu L., W.S. Oetting and D.C. Bennett, 2010. In: European Society for Pigment Cell Research., Available at: http://www.espcr.org/micemut.

Ollmann M.M., M.L. Lamoreux, B.D. Wilson and G.S. Barsh, 1998. Interaction of Agouti protein with the melanocortin 1 receptor *in vitro* and *in vivo*. Genes Development, 12: 316-330.

Powell A.J., M.J. Moss, L.T. Tree, B.L. Roeder, C.L. Carleton, E. Campbell and D.L. Kooyman, 2008. Characterization of the effect of Melanocortin 1 Receptor, a member of the hair color genetic locus, in alpaca (Lama pacos) fleece color differentiation. Small Ruminant Research, 79: 183-187.

Xianzong S. and L.J. Donald, 2006. A new RACE method for extremely GC-rich genes. Analytical Biochemistry, 356: 222-228.

Genetic diversity in Malabari goats

K.A. Bindu[1], K.C. Raghavan[1] and S. Antony[2]
[1]Department of Animal Breeding and Genetics, COVAS, K.A.U., Mannuthy, Thrissur, Kerala, India; binsib@rediffmail.com
[2]Resident Veterinarian, IFCDW, Agadir, Morocco

Abstract

Three hundred goats belonging to Tanur, Thalassery and Badagara regions of the state of Kerala in India were studied for biochemical and microsatellite marker polymorphism, and biometrical traits. A total of eight parameters, viz haemoglobin, transferrin, cerruloplasmin, amylase, albumin, carbonic anhydrase, serum potassium and blood glutathione, were investigated in biochemical polymorphism studies. Of the ten microsatellite primers tested (TGLA 53, INRA 005, INRA 063, ILSTS 005, ILSTS 030, ILSTS 011, HUJ 1177, ETH 10, INHA and BM 720), only three primers were chosen for the study, namely INRA 063,ILSTS 030, and HUJ 1177, which exhibited relatively higher degree of polymorphism. The litter size and body weight of goats were recorded at birth, at one, three, six and twelve months of age and the data analysed. The study revealed that though all populations under investigation had predominant physical characteristics of the Malabari breed, the Tanur population stood apart as regards the biometrical traits, like litter size and body weight in that they were smaller in size and more prolific. The high genetic diversity of the Tanur population was reiterated by microsatellite marker studies.

Keywords: biochemical polymorphism, microsatellite, biometrical traits

Introduction

The domestic goat is alternatively a good source of milk, meat, fibre and skin. The promising prospects of goat-rearing as regards employment generation, financial autonomy and increased living standards of the rural poor can directly translate into poverty alleviation.

The geographically evolved goat breeds display genetic divergence in their production, reproduction, adaptability, heat tolerance and potential for disease resistance. Limited information on the existing diversity in indigenous goat breeds has led to their underutilisation, replacement and dilution through cross breeding. This necessitates estimation and conservation of the genetic variation within breeds.

Goat production in Kerala, India, is centred mainly on its native breed 'Malabari', a dual-purpose goat from North Kerala. The animals have predominant breed characteristics of white and a combination of white with black and brown. They are mostly long-eared and horned with a convex forehead and have rounded udders with funnel-shaped pointed teats. There are significant differences between populations of this breed with regard to traits of economic importance, so the data obtained from any particular population cannot be extrapolated to the breed as a whole. This genetic diversity therefore has to be studied in detail in different populations.

A possible way to study this genetic diversity is by determining genetic variability through polymorphic studies. Polymorphism indicates genetic variability without which there would be no progress made through selection and breeding. This accentuates the need to study polymorphism between breeds as well as within breeds. This study is focused on examining the variations in Malabari goats using biometrical, biochemical and molecular means.

Materials and methods

Three hundred goats of different localities, namely Tanur, Thalassery and Badagara of Kerala state in India, were studied for biochemical and microsatellite marker polymorphism and biometrical traits.

Under biochemical polymorphism a total of eight parameters, viz haemoglobin, transferrin, cerruloplasmin, amylase, albumin, carbonic anhydrase, serum potassium and blood glutathione, were studied.

Haemoglobin was estimated by non-denaturing polyacrylamide gel electrophoresis (PAGE). The whole blood was centrifuged; cell pellets collected and washed with normal saline. Gel slabs were made using eight percent acrylamide gel mix. Tris borate EDTA buffer was used in the top and bottom reservoirs. The cell pellet was diluted with distilled water (1:10). Fifteen microlitres of the diluted cell pellet was loaded in the wells and electrophoresed at 100 V for three hours. The gel was stained with Coomassie Brilliant Blue.

Sodium Dodecyl Sulphate (SDS) PAGE was used for visualising albumin and transferrin bands. Eight percent resolving gel and five percent stacking gel were used. Tris-glycine electrophoresis buffer was used in the top and bottom reservoirs. Fifteen microlitres each of the serum samples were loaded in the wells and electrophoresed at 85 Volts for four hours. The gel was stained with Coomassie Brilliant Blue.

Native PAGE was employed for typing cerruloplasmin, amylase and carbonic anhydrase.

Serum potassium was estimated by flame photometry (Oser, 1965). Malabari goats with a potassium concentration above 22 meq/litre were classified as high potassium type, and those falling below 22 meq/litre were categorised as low potassium type since no bimodality could be observed in the frequency distribution curve.

Glutathione (GSH) level in the whole blood was estimated by the method of Beutler *et al*. (1963), using a spectrophotometer at a wavelength of 412 nm. Goats with GSH values below 60 mg/100 ml RBC were classified as GSH low type and those with values of 60 mg/100 ml RBC and above as GSH high type.

Microsatellite marker study was conducted with the DNA extracted from five millilitres of blood using standard phenol: chloroform extraction procedure. Ten microsatellite primers (TGLA 53, INRA 005, INRA 063, ILSTS 005, ILSTS 030, ILSTS 011, HUJ 1177, ETH 10, INHA and BM 720) were custom synthesised and tested in the present study. Of these, three primers (INRA 063, ILSTS 030 and HUJ 1177), which exhibited a comparatively higher degree of polymorphism, were chosen for the study.

For visualising the PCR products by autoradiography, one of the primers was radiolabelled. The forward primer for each marker was radiolabelled at the 5' end with γ^{32} P-ATP. The reaction was carried out with the DNA End-labeling Kit. The radioactively labeled PCR products were subjected to electrophoresis on six per cent denaturing polyacrylamide gels for better resolution. Denaturing PAGE was performed on the Vertical Sequencer. (Usha, 1995).

The PCR products were mixed with 3.5 µl formamide loading buffer, denatured at 95 °C for 5 min and cooled immediately in ice. About 4µl of this mixture was loaded into each well. Determination of the exact size of alleles necessitated comparison with a sequencing ladder from M13 phage. Sequenced products of M13 DNA, which were also denatured at 95 °C for five minutes, were simultaneously loaded into the middle or side wells. The gels were electrophoresed at 40W for three hours at a constant temperature of around 40 °C. The gels were dried in a gel drier at 80 °C for one and a half hours. After drying the gel was set for autoradiography with X-ray film in a cassette fitted with an intensifying screen. The X-ray film was developed after 24 to 48 hours depending on the intensity of the radioactive signal. The genotypes of animals were determined for each microsatellite loci by comparing the sizes of alleles with an M 13 sequencing ladder.

The genetic distance between the three populations based on allelic frequencies of polymorphic loci of biochemical and microsatellite markers was computed separately using the software POPGENE.

Biometrical traits: the litter size and body weight of goats were recorded at birth, and at one, three, six and twelve months of age and the data was analysed.

Results and discussion

Haemoglobin

All the animals belonging to Tanur and Badagara were of Hb_{AA} type. But the goat population of Thalassery possessed Hb_{AB} and Hb_{BB} phenotypes in addition to Hb_{AA}. Out of 100 animals typed, six animals were of Hb_{AB} and only one was of Hb_{BB} phenotype. The gene frequency of Hb^A and Hb^B was 1 and 0 for Tanur and Badagara populations, while the respective figures were 0.962 and 0.038 for Thalassery population. The frequencies of Hb^A and Hb^B variants in the pooled population were 0.987 and 0.012, respectively indicating a predominance of Hb^A in the population. Similar findings have been reported by Bhat (1985) in Jamunapari goats.

In exotic breeds, a clear predominance of Hb^A variant over Hb^B has been established by Fesus et al. (1983) in Hungarian native goats, by Barbancho et al. (1984) in Spanish goat breeds and by Canatan and Boztepe (2000) and Elmaci (2003) in Turkish goats.

The absence or negligible presence of Hb^B allele in goats, indigenous as well as exotic, may be indicative of either adaptive preference of Hb^A allele to Hb^B allele or species characteristic.

Transferrin

In the goat populations of Tanur, Thalassery, Badagara and pooled population, the gene frequencies of Tf^A observed were 0.995, 1.000, 0.974 and 0.990, respectively indicating the predominance of Tf^A allele in the population. In Thalassery all the animals typed were of Tf_{AA} type while Tanur and Badagara animals had Tf_{AA} and Tf_{AB}. The gene frequencies of Tf^B in Tanur and Badagara goat populations were 0.005 and 0.026, respectively.

The above finding is in agreement with the observations of Fesus et al. (1983) who reported that the majority of the goat breeds in the world have a gene frequency of Tf^A greater than that of Tf^B. Many other workers who also gave the gene frequencies of Tf^A had reported along similar lines while working on various breeds of goats in India, viz. Baruah and Bhat (1980) in Black Bengal goats and Bhat (1987) in Chegu and Changthangi goats.

Similar results in exotic breeds were given by Menrad et al. (1994) in Boer and improved Fawn goats and Canatan and Boztepe (2000) and Elmaci (2003) in hair goats of Turkey.

Albumin, cerruloplasmin, amylase and carbonic anhydrase

No polymorphism was observed for albumin, cerruloplasmin, amylase and carbonic anhydrase. Two bands each were observed for albumin in all the animals studied; a fast-moving band designated as Al [F] and a slow-moving band Al[S], revealing absence of polymorphism at albumin locus.

Two albumin variants in goats had already been reported by Barbancho *et al*. (1984) and Tunon *et al*. (1989) in Spanish goat breeds, and by Vankan and Bell (1992) in Cashmere goats but with higher degree of polymorphism at the locus. In Indian breeds similar findings were reported by Fesus *et al*. (1983) in Beetal, Barbari, their crosses and non-descript natives, by Bhat (1986) in Jamunapari and Sirohi breeds and by Bhat (1987) in Changthangi and Chegu breeds.

The findings of the present work suggests that cerruloplasmin locus is monomorphic, indicating absence of polymorphism as has been reported in other Indian breeds like Beetal, Barbari, Jamunapari, Sirohi, Changthangi and Chegu breeds.

In contrast to the above findings, polymorphism at cerruloplasmin locus was reported by Elmaci (2003) in hair goats of Turkey. But the frequency of the variant allele was very low (0.027).

In Tanur, Thalassery and Badagara goats a single band could be observed for amylase locus indicating the absence of polymorphism. This finding is in agreement with the reports in Indian goat breeds by Bhat (1987) in Changthangi and Chegu goats.

The present work is also in agreement with the studies conducted in exotic breeds by Morera *et al*. (1983) in Spanish Merino sheep, by Shamsuddin *et al*. (1986) in Saanen and Alpine halfbreeds with Malabari, by Tunon *et al*. (1989) in Spanish goats and by Menrad *et al*. (1994) in German improved Fawn and Boer goats.

In contrast to the above findings, polymorphism for amylase locus was observed in Turkish hair goats (Elmaci, 2003). Two variants for amylase locus were observed but the frequency of the variant allele was very low.

Serum potassium

In the present study the potassium concentration ranged between 1.3 and 15.7 meq/l, all indicative of low potassium types. Maximum concentration of serum potassium was detected in Badagara population while Tanur animals possessed the lowest. The concentrations of serum potassium in Tanur, Thalassery and Badagara were 3.45±0.11, 4.52±0.20 and 4.60±0.14 meq/l, respectively. Least square analysis of variance of potassium concentration in different populations showed that there is a significant difference between goat populations. The low potassium types might have got fixed in the population and may be due to some linkage to some traits of economic importance or some adaptive importance.

Absence of polymorphism has also been reported by Bhat *et al*. (1983) in Jamunapari and Barbari herds, who opined that unlike in sheep, not all goat breeds may exhibit polymorphism at potassium locus. So also, Bhat (1986) reported that in the Jamunapari breed, only HK type could be detected which got fixed during the course of evolution of this breed. Furthermore, studies in Pashmina goats by Bhat and Singh (1987) agreed with the above finding, where only in HK type there was an absence of polymorphism at potassium locus, with a mean serum potassium concentration of 29.05 meq/l.

Glutathione (GSH)

The present study pointed to a predominance of the low GSH type on the whole. The low GSH type animals (49%) existed in almost equal proportions to the high GSH type in Badagara, while in Tanur, the low GSH type animals were in lesser proportions (30%). The Thalassery population, however, had a marked predominance of the low GSH type (83%), establishing a predominance of low GSH type (53.67%) in the pooled population in the final analysis. The gene frequencies of GSH^H were 0.45, 0.09 and 0.30 in Tanur, Thalassery and Badagara goat populations, respectively, whereas the respective gene frequencies of GSH^h were 0.55, 0.91 and 0.70. Least square analysis of variance of GSH values showed that there was a significant difference between the different subpopulations. More et al. (1980) reported that Jamunapari animals had relatively more numbers of low GSH animals.

In goats of Badagara, a place of geographical proximity to Thalassery, more or less equal proportions of GSH high (51%) to GSH low types (49%) were detected with the GSH high type gaining only a slight edge. The reason for this may be attributed to the relatively close confinement of these two populations. Moreover, a process of natural selection, on the grounds of economic traits possibly having some linkage to the gene for low glutathione type, might be going on in this breeding tract resulting in a predominance of low glutathione type in the populations as a whole. Furthermore, it should be noted in this context that Malbari goats also have Jamunapari blood in which the GSH low types predominates, as established already. Tunon et al. (1989) had however reported the absence of polymorphism with regard to GSH loci in Spanish goats.

Genetic distance

In the present study, genetic distances between different populations were computed using Nei's method (1978), based on polymorphic loci, viz. haemoglobin, transferrin and glutathione. Genetic distance between Tanur and Thalassery was 0.0498, and between Tanur and Badagara 0.0092, whereas between Badagara and Thalassery it was 0.0166. In this study the maximum genetic distance found between Tanur and Thalassery population was in agreement with the geographical distance between the populations.

The data obtained from studies of blood genetic systems should be combined with other data for more accurate interpretations as regards phylogenetic relationships.

Pepin and Nguyen (1994) calculated genetic distances based on the variation in allelic frequencies between breeds, and concluded that the results were in close agreement with historical and geographical data of the breeds examined. Nguyen et al. (1992) opined that the observed genetic differences between Rambouillet and Spanish Merino could be attributed to the evolutionary change due to random drift in the small and closed flock of Rambouillet.

Tunon et al. (1989) calculated the genetic distances between 14 Spanish goat breeds in a similar way using Nei's distance and opined that the distance values ranged from 0.003 to 0.097.

Nei (1978) emphasised that the dendrogram only represents the genetic relationship between breeds, but may or may not show the true evolutionary history of populations, especially when they are not completely isolated.

Protein polymorphic markers being based on a limited number of expressed genes, the genetic distance calculated by this method fails to give a conclusive picture. The only possible conclusion that could be arrived at from the above studies is the existence of a relationship between these populations. It further suggests that the Tanur population is distinctly different from the other two populations. Hence an attempt was made to study the different populations at the molecular level, using microsatellite markers.

The two clear advantages of microsatellites over other DNA markers is that they have multiple alleles as well as high heterozygosity frequencies, which make them highly informative for genetic analysis (Gill et al.,1994). The values of genetic distance obtained from microsatellite data might be more precise when compared to that obtained from protein markers (Arranz et al., 1996).

Microsatellite markers

Microsatellites have proved to be useful polymorphic markers for the analysis of genetic relationships. The usefulness of microsatellite markers for the estimation of genetic distances among closely related populations has been documented in numerous studies (Ciampolini et al., 1995 and Arranz et al., 1996).

Ten microsatellite primers (TGLA 53, INRA005, INRA 063, ILSTS 005, ILSTS 030, ILSTS 011, HUJ 1177, ETH 10, INHA and BM 720) were tested in the present study. Out of the ten microsatellite markers tested, seven were monomorphic in goats. Three primers (INRA 063, ILSTS 030 and HUJ 1177) were found to be polymorphic (Figures 1, 2 and 3). A total of 16 genotypes were observed for INRA 063 in Tanur and 11 each in Thalassery and Badagara populations, respectively. The maximum number of alleles (10) was observed in Tanur population while the minimum (5) in Thalassery. Badagara recorded eight alleles. Six alleles (163/155, 163/159,

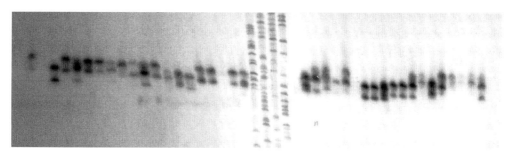

Figure 1. Autoradiograph showing polymorphism at INRA 063 locus.

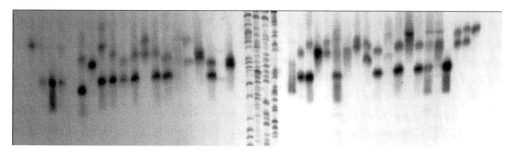

Figure 2. Autoradiograph showing Polymorphism at HUJ 1177 locus.

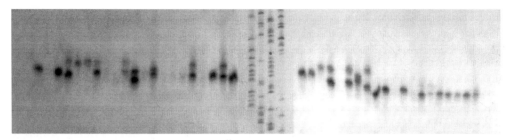

Figure 3. Autoradiograph showing polymorphism at ILSTS 030.

165/157, 169/163, 173/163 and 173/165) were specific to Tanur population, while three alleles each were specific to the other two populations (163/163, 165/159 and 171/171 in Thalassery and 167/165,169/157and 169/167 in Badagara). The maximum values for heterozygosity (0.847) and PIC (0.843) were observed in Tanur while the lowest were recorded in Thalassery (0.656 and 0.652, respectively). In Badagara, values for heterozygosity and Polymorphic Information Content (PIC) were 0.826 and 0.815, respectively. Chenyambuga et al.(2004) reported ten alleles with a size range of 141 to 179 bp in goats of Subsaharan Asia.

The maximum number of alleles and genotypes for HUJ 1177 was found in the Tanur population. The number of alleles observed in Tanur, Thalassery and Badagara populations were 15, 14 and 13, respectively. Twenty-seven genotypes were recorded in goat populations of Tanur, while 24 each were recorded in Thalassery and Badagara populations. Both values for heterozygosity and PIC in Tanur and Badagara were recorded as 0.894 and 0.880, while at Thalassery the heterozygosity and PIC values stood at 0.909 and 0.907, respectively.

The maximum number of alleles at ILSTS 030 locus was observed in Tanur (11), whereas the minimum was recorded in Thalassery (7) while Badagara recorded nine. The highest values for heterozygosity (0.850) and PIC (0.848) were detected in the Tanur population, while the lowest (0.793 and 0.775) were recorded in Thalassery. The heterozygosity and PIC values were 0.828 and 0.823, respectively, in Badagara. Allele size range at this locus in Marwari goat breed was reported to be 164 to 174 bp (Kumar et al., 2005).

The highest mean number of alleles per locus as well as high allelic size range was noticed in the Tanur goat population. Hence Tanur goat population can be considered to be more genetically diverse than the other groups based on the number of alleles and size range.

Genetic distance computation based on the allelic frequencies of microsatellite markers revealed that Tanur and Badagara populations were the most separated, while a close relationship was suggested between the Thalassery and Badagara populations. The genetic distance between Thalassery and Badagara was found to be 0.5729, between Thalassery and Tanur 0.7795 and between Badagara and Tanur 0.8401.

Genetic distances calculated (Nei's formula, 1978) from microsatellite markers are 62.27-fold greater than from protein markers. The latter gave values ranging from 0.0092 to 0.0498, whereas distances based on microsatellite data varied from 0.57 to 0.84.

Arranz et al. (1996) observed that genetic distance calculated using the microsatellite marker system is usually higher when compared to the other marker systems because of greater variation at these loci. The present findings also support the observations of Arranz et al. (1996).

The dendrogram of relationship between the three different goat populations showed one cluster, grouping Thalassery and Badagara populations (Figure 4). From the genetic distance values calculated, it is evident that genetic distance between Tanur and Thalassery and Tanur and Badagara is greater than between Thalassery and Badagara, which is in accordance with the geographical distribution of the breedable area. This can be substantiated by the fact that because of the geographical proximity, close breeding may be occurring between Thalassery and Badagara populations and they are found to be closely related. Tanur population is different from the other two goat populations.

Biometrical traits

The Tanur animals had a higher litter size with more incidences of twins, triplets and quadruplets in comparison to the other populations studied. The body weight was minimum in the Tanur population while Thalassery animals recorded the maximum. The mean birth weights recorded in kilogram in Tanur, Thalassery and Badagara were 1.73±0.04, 1.88±0.03 and 1.86±0.05 kg, respectively. The body weights in the first year of life were 18.37±2.07, 27.00±1.75 and 25.58±0.76 kg, respectively (Table 1).

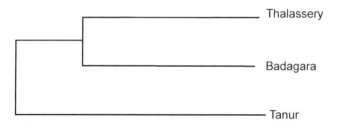

Figure 4. Dendrogram based on genetic distance.

Table 1. Body weight (kg) of Malabari goats.

Populations	Birth weight	One month weight	Three months weight	Six months weight	One year weight
Tanur	1.73±0.04	4.42±0.04	8.83±0.30	17.83±1.54	18.37±2.07
Thalassery	1.88±0.03	5.37±0.17	10.02±0.24	21.00±0.47	27.00±1.75
Badagara	1.86±0.05	4.50±0.15	9.52±0.26	21.36±0.55	25.58±0.76

Conclusion

The techniques employed here could be used effectively to differentiate breeds and different populations of the same breed. Of these, microsatellite markers proved to be the most effective one. Genetic distance computation based on biochemical and microsatellite marker studies revealed that the Tanur animals remained most separated from the other two populations. The observations on biometrical traits reiterated the finding that Tanur animals were different, though all three populations possess the characteristics of the Malabari breed. The existence of high genetic variability of different populations of the same breed is actually a gift, the possibilities of which have been hitherto underutilised in breeding. The identification of uniqueness in the Tanur population in many respects highlights the need for further similar investigations in case of other goat populations and helps to formulate an effective selection programme. While conserving animals, genetic variability needs to be studied so that the desirable traits could be given due importance in further propagation. Ultimate benefactors would be the rural poor, who have the most to gain when provided with better animals. The lower body weight of Tanur animals is welcome in that even women can manage them, while their high prolificacy gives added value as regards economy, the most important concern of the average Keralite marginal farmer. In a nutshell, the study asserts the need for effective utilisation of Tanur animals in evolving smaller goats with higher prolificacy to attain a beneficial end result for the economically backward and under-privileged farmers.

References

Arranz J.J., Y. Bayon and F. San Primitivo, 1996. Comparison of protein markers and microsatellites in differentiation of cattle populations. Animal Genetics, 27: 415-419.

Barbancho M., D. Llanes, L. Morera, R. Garzon and A. Rodero, 1984. Genetic markers in the blood of Spanish goat breeds. Animal Blood Groups Biochemical Genetics, 15: 207-212.

Baruah P. and P.P. Bhat, 1980. Note on the genetics of haemoglobin and transferrin polymorphisms in three breeds of Indian goats. Indian Journal of Animal Science, 50: 576-579.

Beutler E., O. Duron and B.N. Kelly, 1963. Improved method for the determination of blood glutathione. Journal of Laboratory Clinical Medicine, 61: 882-888.

Bhat P.P., 1985. Genetic markers in Jamunapari and Sirohi goat breeds. Indian Journal of Animal Science, 55: 447-452.

Bhat P.P., 1986. Genetic markers in Jamunapari and Sirohi goat breeds. Indian Journal of Animal Science, 56: 430-433.

Bhat P.P., 1987. Genetic studies on biochemical polymorphisms of blood serum proteins and enzymes in Pashmina goats. Indian Journal of Animal Science, 57: 598-600.

Bhat P.P. and V.P. Singh, 1987. Alkaline phosphatase, whole blood potassium, sodium and per cent packed cell volume in Pashmina goats. Indian Journal of Animal Science, 57: 773-774.

Bhat P.P., T.C. Santiago and N.K. Sinha, 1983. Blood potassium, sodium, haemoglobin and transferrin polymorphism in Jamunapari and Barbari goats. Indian Journal of Animal Science, 53: 1151-1152.

Canatan T. and S. Boztepe, 2000. The polymorphisms of haemoglobin and transferrin in Turkish hair goat. Indian Veterinary Journal, 77: 966- 968.

Chenyambuga S.W., O. Hanotte, P.C. Watts, S.J. Kemp, G.C. Kifaro, P.S. Gwakisa, P.H. Petersen and J.E.O. Rege, 2004. Genetic characterization of indigenous goats of south Saharan Africa using microsatellite DNA markers. Asian-Australian Journal of Animal Science, 17: 445-452.

Ciampolini R., K. Goudarz, D. Vaiman, J.L. Foulley and D. Cianci, 1995. Individual multilocus genotypes using microsattelite polymorphisms to permit the analysis of genetic variability within and between Italian beef cattle breeds. Journal of Animal Science, 73: 3259-3268.

Elmaci C., 2003. Some genetic markers in the native hair goats of Turkey. Indian Veterinary Journal, 80: 223-225.

Fesus L., S. Varkonyi and A. Ats, 1983. Biochemical polymorphisms in goats with special reference to the Hungarian Native breed. Animal Blood Groups Biochemical Genetics, 14: 1-6.

Breeding and genetics

Gill P., P.L. Ivanov, C. Kimpton, R. Piercy, N. Benson, G. Tally, I. Evett, E. Hagelberg and K. Sullivan, 1994. Identification of the remains of Romanov family by DNA analysis. Nature Genetics, 6: 130.

Kumar D., S.P. Dixit, R. Sharma, K. Pandey, G. Sirohi, A.K. Patel, R.A.K. Agarwal, N.K. Verma, D.S. Gour and S.P.S Ahlawat, 2005. Population structure, genetic variation and management of Marwari goats. Small Ruminant Research, 9: 41-48.

Menrad M., E. Muller, C.H. Stier and C. Geldermann Gall, 1994. Protien polymorphisms in the blood of German Improved Fawn and Boer goats. Small Ruminant Research, 14: 49-54.

More T., A.K. Rai and M. Singh, 1980. Red cell glutathione polymorphism in goats. Indian Journal of Animal Science, 50: 1012-1014.

Morera L., D. Llanes, M. Barbancho and A. Rodero, 1983. Genetic polymorphism in Spanish Merino sheep. Animal Blood Groups Biochemical Genetics, 14: 77-82.

Nei M., 1978. Estimation of average heterozygosity and genetic distance from a small number of individuals. Genetics, 89: 583-590.

Nguyen T.C., L. Morera, D. Llanes and P. Leger, 1992. Sheep blood polymorphism and genetic divergence between French Rambouillet and Spanish Merino: role of genetic drift. Animal Genetics, 23: 325-332.

Oser B.L., 1965. Hawk's Physiological Chemistry. Fourteenth edition. Mc Graw-Hill Book Company, New York, NY, USA, p. 650.

Pepin L. and T.C. Nguyen, 1994. Blood groups and protein polymorphisms in five goat breeds (Capra hircus). Animal Genetics, 25: 333-336.

Shamsuddin A.K., B. Nandakumaran and G. Mukundan, 1986. Genetic studies on haemoglobin albumin and amylase polymorphism in Malabari goats and its exotic cross breds. Kerala Journal of Veterinary Science, 17: 1-6.

Tunon M.J., P. Gonzalez and M. Vallejo, 1989. Genetic relationships between 14 native Spanish breeds of goat. Animal Genetics, 20: 205- 212.

Usha A.P., 1995. Microsatellite markers in genetic improvement of livestock. Ph.D thesis, University of Edinburgh, Edinburgh, UK. p. 171.

Vankan D.M. and K. Bell, 1992. Genetic polymorphism of plasma vitamin D – binding protein (GC) in Australian goats. Animal Genetics, 23: 457-462.

Genetic parameters for growth of fibre diameter in alpacas

L. Varona[1], I. Cervantes[2], M.A. Pérez-Cabal[2], R. Morante[3], A. Burgos[3] and J.P. Gutiérrez[2]
[1]*Unidad de Genética Cuantitativa y Mejora Animal. Universidad de Zaragoza, Spain*
[2]*Departamento de Producción Animal, Universidad Complutense de Madrid, Avda. Puerta de Hierro s/n, E-28040 Madrid, Spain; gutgar@vet.ucm.es*
[3]*Pacomarca S.A., P.O. BOX 94, Av. Parra 324, Arequipa, Peru*

Abstract

Alpaca is the most important fibre producer of South American camelid species, and an important income for the Andean communities. Nowadays, the fibre diameter is considered the main selection criterion in alpaca populations all over the world. However, fibre diameter increases with the age of the animals, and it would be preferable to select those animals that maintain a thin fibre throughout their life. The aim of this study is to describe the genetic relationship between fibre diameter at birth and its evolution over a lifetime. The analysis of the evolution of fibre diameter was studied as a longitudinal trait using the Bayesian procedure for the analysis of production functions. The results suggested that there is substantial genetic variation in fibre diameter at birth and also on the linear growth of fibre diameter. This confirms the need for a genetic programme to modify the evolution of the fibre diameter over time. However, selection to increase the growth of the fibre without a substantial reduction in the fibre diameter at birth seems implausible.

Keywords: alpaca, evolution, fibre diameter

Introduction

Alpaca is the most important fibre producer of South American camelid species, and an important income for the Andean communities. There are two alpaca breeds with different fleece characteristics: Huacaya (HU), which represents 85% of the alpaca population in Peru (Quispe *et al.*, 2009) and Suri (SU).

Genetic programmes for fibre characteristics have been implemented by PACOMARCA S.A. by using a textile value index as selection criterion. In addition, estimates of heritabilities for fibre traits suggest moderate genetic determinism (Cervantes *et al.*, 2010).

Fibre diameter is nowadays considered the main selection criterion in alpaca populations all over the world (Gutiérrez *et al.*, 2009). However, fibre diameter increases with the age of the animals, and it would be preferable to select those animals that maintain a thin fibre throughout their life.

The analysis of the evolution of fibre diameter can be studied as a longitudinal trait using the Bayesian procedure for the analysis of production functions described by Varona *et al.* (1997). Thus, the aim of this study is to describe the genetic relationship between fibre diameter at birth and its evolution over a lifetime.

Material and methods

The data set consists of a pedigree of 3,621 individuals of the Huacaya breed, and 6,808 records of fibre diameter corresponding to 2,784 individuals. The average fibre diameter was 23.01

μm and standard deviation was 4.17 μm. The relationship between age and fibre diameter is presented in Figure 1.

Statistical model

The statistical model assumed that the jth measure of the fibre diameter for the ith individual (y_{ij}) is determined by the effect of the month-year of recording (f_i with 36 levels), the fibre diameter at birth (a_i) and the slope of linear growth for fibre diameter (g_i) time the age at the jth measure (x_{ij}) plus a residual (r_{ijk}):

$$y_{ijk} = f_i + a_j + g_j x_{jk} + r_{ijk} \tag{1}$$

The residuals (**r**) are assumed to be Gaussian and identically distributed:

$$\mathbf{r} \sim N(\mathbf{0}, \mathbf{I}\sigma_r^2) \tag{2}$$

where σ_r^2 is the residual variance.

Fibre diameter at birth and the slope of the increase in fibre diameter are assumed to be determined by systematic (sex -2 levels- and colour -3 levels-), genetic and environmental effects.

$$\begin{bmatrix} \mathbf{a} \\ \mathbf{g} \end{bmatrix} = \begin{bmatrix} \mathbf{X} & \mathbf{0} \\ \mathbf{0} & \mathbf{X} \end{bmatrix} \begin{bmatrix} \mathbf{b}_a \\ \mathbf{b}_g \end{bmatrix} + \begin{bmatrix} \mathbf{Z} & \mathbf{0} \\ \mathbf{0} & \mathbf{Z} \end{bmatrix} \begin{bmatrix} \mathbf{u}_a \\ \mathbf{u}_g \end{bmatrix} + \begin{bmatrix} \mathbf{e}_a \\ \mathbf{e}_g \end{bmatrix} \tag{3}$$

The additive genetic effects and the environmental effects were assumed to be distributed with the following multivariate Gaussian distributions:

$$\begin{bmatrix} \mathbf{u}_a \\ \mathbf{u}_g \end{bmatrix} = N(\mathbf{0}, \mathbf{A} \otimes \mathbf{G}) \tag{4}$$

$$\begin{bmatrix} \mathbf{e}_a \\ \mathbf{e}_g \end{bmatrix} = N(\mathbf{0}, \mathbf{A} \otimes \mathbf{G}) \tag{5}$$

Where **G** and **R** are the matrices of additive genetic and environmental (co)variances components. Prior distributions for $\mathbf{b}_a$, $\mathbf{b}_g$, **G**, **R** and σ_r^2 were assumed to be flat between bounded limits.

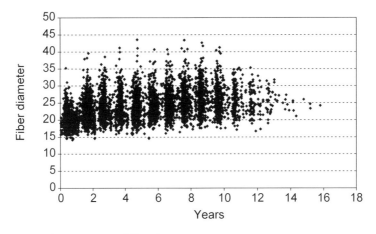

Figure 1. Relationship between age and fibre diameter.

The Bayesian analysis was implemented with a Gibbs sampler algorithm with a single long chain of 1,000,000 iterations after discarding the first 250,000.

Results and discussion

The results of the posterior distributions of the variance components are presented in Table 1.

Moreover, the posterior mean estimate of the residual variance was 5.125 with a posterior standard deviation of 0.115. These results suggest that there is substantial genetic variation in fibre diameter at birth and also in the linear growth of fibre diameter. The last results imply that a genetic programme is plausible for modifying the evolution of the fibre diameter over time. However, the posterior mean estimate of the additive genetic correlation between diameter at birth and linear growth was 0.889 with a posterior standard deviation of 0.105. This result is confirmed by the plot of the posterior mean estimates of the breeding values for both traits (Figure 2).

Table 1. Posterior mean and standard deviation (SD) of genetic parameters of the fibre diameter at birth (a) and its posterior linear growth (c).

	Diameter at birth (a)		Linear growth (c)	
	mean	SD	mean	SD
σ^2_u	1.212	0.161	0.052	0.011
σ^2_e	0.142	0.062	0.134	0.018

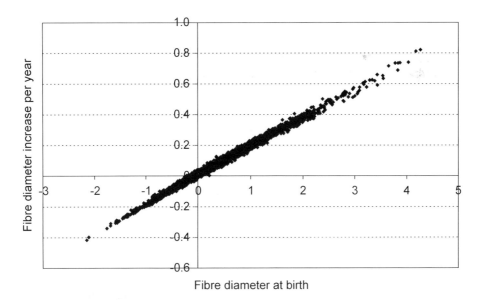

Figure 2. Posterior mean estimate for breeding values for fibre diameter at birth and linear growth.

Therefore, selection to increase the growth of the fibre without a substantial reduction in the fibre diameter at birth seems implausible. Further research is needed.

References

Cervantes I., M.A. Pérez-Cabal, R. Morante, A. Burgos, C. Salgado, B. Nieto, F. Goyache and J.P. Gutiérrez, 2010. Genetic parameters and relationships between fibre and type traits in two breeds of Peruvian alpacas. Small Ruminant Research, 88: 6-11.

Gutiérrez J.P., F. Goyache, A. Burgos and I. Cervantes, 2009.Genetic analysis of six production traits in Peruvian alpacas. Livestock Science, 123: 193-197.

Varona L., C. Moreno, L.A. García-Cortés and J. Altarriba, 1997. Multiple trait genetic analysis of underlying biological variables of production functions. Livestock Production Science, 47: 201-209.

Quispe E.C., T.C. Rodríguez, L.R. Iñiguez and J.P. Mueller, 2009. Producción de fibra de alpaca, llama, vicuña y guanaco en Sudamérica. Animal Genetic Resources Information, 45: 1-14.

Comparison of different breeding strategies to improve alpaca fibre production in a low-input production system

E.C. Quispe[1], R. Paúcar[1], A. Poma[1], A. Flores[2] and L. Alfonso[3]
[1]Improvement Programme for South American Camelids, National University of Huancavelica (UNH)- Huancavelica, Peru; edgarquispe62@yahoo.com
[2]Graduate School, National Agrarian University the Molina (UNALM), Lima, Peru
[3]Animal Production Area, Public University of Navarra, Spain

Abstract

Alpaca production in Huancavelica (Peru) is a low-input production system characterised by a high number of small herds with feeding based on the use of natural resources. Producer's income depends basically on the quantity (unwashed fleece weight, UFW) and the quality (fibre diameter, FDM) of the fibre commercialised. Productive and genealogical information is difficult and expensive to record because of the geographical characteristics (altitude>3,800 m) and the lack of effective communication between communities of alpaca's owners. The aim of this study was to analyse different breeding schemes in terms of genetic response and inbreeding level, but also considering cost and participation of farmers. A genotypic aggregate value defined as a linear function of UFW and FDM was considered as breeding goal. Six different strategies were considered, involving different numbers of selected males in a similar way to Wurzinger et al. (2009) in Bolivian llamas: three considering one central nucleus with artificial insemination (15 males -NAI-) or natural mating (50 males -NM50-, 33 males -NM33-), two considering a communal nucleus involving 23 herds (79 males -C79-, 113 males -C113-) and a combination of both on the basis of natural mating, one central nucleus and a communal nucleus involving 12 herds (81 males -NC-). SelAction software (Rutten et al., 2002) was used to derive genetic response and inbreeding level assuming different values for phenotypic and genetic parameters of UFW and FDM, and different amounts of phenotypic information. With respect to the C113 scheme, NAI increases the genetic response in a 50%, NM33 in a 30%, NM50 in a 20%, and C79 and NC in a 10%. In average, the increase of inbreeding was 0.90 % per generation for NAI, 0.35% for NM33, 0.20% for NM50, 0.10% for C79 and NC, and 0.05% for C113. Communal nucleus schemes, alone or combined with a central nucleus, showed a low rate of genetic response but they could be useful to start a breeding programme in Huancavelica. That is because of the lower inbreeding level, the lower economic costs, and the higher level of participation of farmers as opposed to central nucleus schemes only, especially when they make use of artificial insemination.

Keywords: Alpaca, breeding, scheme, response, inbreeding

Introduction

A genetic improvement plan is subject to completion of a four-step sequence: determining the type of animal to breed or the desired improvement; choosing the information or the selection criteria to be used or applied; selecting the animals based on their phenotypic parameters or the genetic merit of the relevant characters; and designing the mating process for the animals selected to ensure a suitable diffusion of genetic material (Kosgey et al., 2006).

The choice of alpacas – as referred to by Mason and Buvanembra (1982) with regards to local cattle breeds – has some arguments in its favour, since: (1) they can adapt to living within their environment, showing a resilience to illness, they are able to survive and breed under harsh geographical and climate conditions, and they make the most out of low quality food and restricted water supply; (2) the increase in their genetic merit enables them to not only minimise competition from other exotic species, but also constitutes the main and most effective means to fight poverty and food insecurity in the farming communities of the South American countries they inhabit (Quispe *et al.*, 2009b).

In large-scale production of small ruminants, the inclusion of them all into the active part of the selection programme is neither valid nor practical, since it poses difficulties in ensuring adequate control and brings high costs of measurement and registration (Kosgey and Okeyo, 2007). In addition, in the case of flocks and herds of animals present in developing countries, where resources are scarce, it is appropriate to optimise however little is available, which is possible when genetic improvement is generated in a small portion of the population (referred to as nucleus), while also improving controls. Performance recording and genetic evaluation are carried out on the nucleus and genetic progress is spread to the commercial population by using a breeding male and/or semen (whenever artificial insemination is feasible) (Kosgey *et al.*, 2006).

A good starting point in designing selection schemes within a genetic improvement programme is to evaluate similar improvement programmes that have or are being conducted and, in this way, genetic improvement opportunities and options can be properly understood. In developing countries where breeders have relatively small herds – as is the case of alpaca production in Peru and Bolivia – despite the great effort made, only slight genetic progress has been attained, mainly due to institutional, technical and infrastructure issues (Kosgey *et al.*, 2006). These not only determine the most appropriate selection scheme to be implemented, but they also have to do with sustainability, which is indispensable in the genetic improvement process since it involves much time and results are usually seen within or after some years.

It has been generally suggested that the best selection scheme for developing countries seems to be a centralised nucleus herd programme, where the nucleus is located at the heart of a state-managed production centre (Gizaw *et al.*, 2009). The advantage of maintaining a centralised reproduction nucleus is that it ensures accuracy in data recording and processing, which requires the involvement of experts to conduct various activities on the nucleus (which would lead to the breeder being excluded). Therefore, selection can be carried out efficiently to achieve high genetic gains. However, this can only work for as long as there is financial and technical support (whether it be from the government or NGOs) and as long as there is willingness, empowerment and commitment from the individual breeders, which means that long-term use would not be sustainable. Another alternative would be the dispersed nucleus scheme, in which several breeders form a nucleus. For this to happen, it is necessary for farmers to be well trained and willing to keep production and pedigree records. Hence, the genetically improved animals may subsequently be distributed to other breeders and, as a manner of compensation, the beneficiaries must give the first offspring born from the reproduction animals provided back to the dispersed nucleus (Wurzinger *et al.*, 2008). In this case, it is still necessary to have a data processing centre, although there is greater involvement by the breeders. However, the lack of technicians, infrastructure and proper organisation thereof would mean this system would be overly ambitious and, although some research has been done on tropical sheep (Gizaw *et al.*, 2009) and llamas (Wurzinger *et al.*, 2008), no such information has been obtained on alpacas.

Gizaw *et al.* (2009) proposed and evaluated a centralised communal scheme for ovine species, based on the active participation of the farmers within the community, which entails the formation of a nucleus in the community through the selection of better-quality sheep owned by the farmers taking part. This could be more acceptable within the community, though it must be highlighted that in this case the mass selection based on individual evaluations of phenotypes with a lack of pedigree records would be necessary. In this system, breeders and experts work together on collecting data, evaluating the potential reproduction candidates and selecting the animals. However, the operational aspects are also indicated as being more complex, and therefore, said scheme is prone to causing issues with the implementation of record-keeping and selection methods.

In a simulation study on alpaca herds, León-Velarde and Guerrero (2001), found that the use of a selection index generates gains of 300 g, 6 cm and -1.4 (which represents 30 g, 0.6 cm and 0.14/year) for unwashed fleece weight, staple length and fibre diameter, on a 10-year selection. In addition, Ponzoni (2000) considered in a simulation study for Australian herds that using a compound selection index would enable a genetic gain of 2.50 to 3.36 kg, 25.0 to 22.2 µm and 63.0 to 63.7 kg for washed fleece, fibre diameter and liveweight gain, respectively, over a 10-year period, which represents an annual genetic progress of 86 gr, -0.28 µm and 70 g. for the abovementioned features. Quispe *et al.* (2007), predicting a genetic gain per selection under high Andean production systems characterised by low resources, bearing in mind the weight of unwashed fleece and fibre diameter, reported an annual progress of 77.5 g and -0.07 µm, respectively. However, Quispe *et al.* (2009a) reported gains per year ranging from 49.77 to 103.25 gr and -0.08 to -0.20 µm, respectively, for the abovementioned characters, while under the same conditions.

Based on these considerations, the study herein was carried out with the aim of analysing various selection schemes in terms of genetic progress, taking into account the cost and involvement of the alpaca breeders; therefore, the aggregate genotype value was considered in the breeding objective, including unwashed fleece weight and fibre diameter based on a linear function. The increase in inbreeding as a result of the selection process is also considered.

Materials and methods

The study was conducted on the Huancavelica region, which comprises areas that are between 4,000 and 4,800 metres above sea level, with temperatures ranging from -5 °C to 0 °C at night and between 14 °C and 18 °C during the day, with rainfall reaching 752.4 mm/year.

Prior to proposing the improvement schemes for white Huacaya alpacas, a division was made of the 225,000 alpacas that have been bred in Huancavelica (Quispe, 2005) based on fleece colour, age range and improvement level, in accordance with the methods proposed by Quispe *et al.* (2009a) in order to arrange the animal data into 3 stratified levels: (1) selection nucleus; (2) multiplier herds; and (3) penned herds, for which production and reproductive indexes were used (Table 1) based on bibliographical reviews (Nieto and Alejos, 1999; Brenes *et al.*, 2001; Wurzinger *et al.*, 2008; Quispe *et al.*, 2009a; Quispe *et al.*, 2009b).

A selection nucleus was proposed made up of 1,500 females and 50 to 113 males, which enables us to produce a sufficient number of breeding specimens to replace males from two population strata (selection nucleus and multiplier herds). The proposed number of 1,500 females corresponds to 23 herds found in Huancavelica by the PROCASUD (South American Camelid Improvement Programme), with which it has been working, each herd containing an average of 64 females for breeding.

By analysing the production system and taking into account the logistics and infrastructure that are most important for the effectiveness and sustainability of any genetic improvement programme (Kosgey and Okeyo, 2007), three types of breeding schemes are proposed: (1) A scheme within and across production units, where evaluation and selection of animals for

Table 1. Production and reproductive indexes for the stratification of the alpaca population in the Huancavelica region.

Indexes	Amount
Alpaca population in Huancavelica	225,000
No. of females per breeding male	30-45
Breeding male's reproductive lifespan (in years)	4
% of breeding males for annual stud replacement	25
% of coloured alpacas	14
Birth rate	45
Mortality rate (offspring)	15
Mortality rate (adults)	6
% of males out of offspring born	50
% of adult animals with genetic body and fleece defects	20
% of animals in white alpaca pen/population	95
Average number of white alpacas per herd	120
Miscellaneous expenses per kg processed fibre (in *nuevos soles*)	9.94
Makeup of the white alpaca population based on records (%)	
Offspring (male and female)	18.0
Young alpacas (< 3 years)	25.0
Breeding males	4.0
Breeding females	53.0

replacement purposes are carried out within the production units, with and without reduction of up to 50% of breeding males; (2) A scheme based on a partial reproductive nucleus where reproductive specimens provided on loan by the production units are maintained at 50%, and the other 50% is generated from selection within each production unit; and (3) A scheme with a reproductive specimen nucleus, at a proportion of three females per male [1:30, 1:45 and 1:100 (Table 2)]. In both schemes, the scenarios where productive record-keeping was conducted on males and females were taken into account, as well as the scenario where productive and pedigree record-keeping was used.

The number of males for potential selection produced by the breeding females from the selection nucleus was determined according to the methods described by Wurzinger *et al.* (2008), as shown in Figure 1.

The evaluation of the proposed selection schemes, considering for each of them the 3 forms of identifying the best animals, according to the type of information and evaluation applied (IP = selection using the specimen's individual performance; IP+BLUP = selection using the BLUP method with information on the specimen's performance; and IP+BLUP+S = selection using the BLUP method with information on the specimen and related specimens evaluated) and under 3 levels of genetic correlation between UFW and FDM (low = 0.12; medium = 0.25; and high = 0.50), was performed in terms of genetic progress and the inbreeding coefficient, those estimated using the SelAction programme (Rutten and Bijma, 2001), for which estimated genetic parameters were used in this study (economic weighting, phenotypic variance and phenotypic correlation) and genetic heritability and correlation as obtained by Gutiérrez *et al.* (2009), which are shown in Table 3.

Table 2. Selection schemes and scenarios considered in terms of herds, males selected and males on loan.

Schemes	No. of herds	No. of studs needed for mating	No. of studs needed as replacements per year	Percentage of studs selected
Breeding specimen nucleus				
Scenario 1.1.	1	50	13	0.0579
Regular handling of breeding specimens				
Scenario 1.2.	1	33	8	0.0386
Good handling of breeding specimens				
Scenario 1.3.	1	15	4	0.0174
Artificial insemination				
Breeding specimen nucleus and farmers' herds				
Scenario 2.1.	1	25	20	0.0944
Breeding specimens nucleus				
Farmers' herds	12	56		
Farmers' herds only				
Scenario 3.1.	23	113	28	0.1308
Without reduction of breeding males				
Scenario 3.2.	23	79	20	0.0916
With a 30% reduction of breeding males				

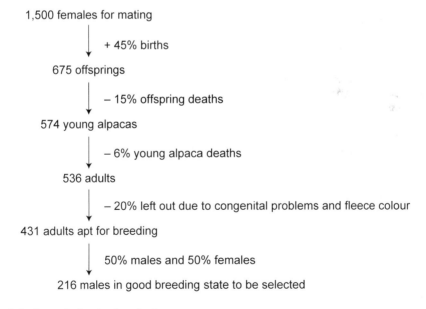

1,500 females for mating

↓ + 45% births

675 offsprings

↓ − 15% offspring deaths

574 young alpacas

↓ − 6% young alpaca deaths

536 adults

↓ − 20% left out due to congenital problems and fleece colour

431 adults apt for breeding

↓ 50% males and 50% females

216 males in good breeding state to be selected

Figure 1. Scaling to find males for selection.

Table 3. Parameters applied to determine genetic progress.

Parameters	FDM (mic)	UFW (g)
Phenotypic variance	9.23^2	$279,744.17^2$
Economic value (a)	-4.0000^2	0.0240^2
Repeatability	0.57^1	0.32^1
Heritability, phenotypic	0.41^1	0.12^2
genetic, heritability	0.12^1	0.25^3
R common environment	0.62^1	

[1] Genetic parameters taken from Gutiérrez et al. (2009).
[2] Phenotypic parameters obtained in this study.
[3] Parameters taken from Quispe et al. (2009a).

Results and discussion

Proposed selection schemes

Six selection strategies were proposed, based on the farmers' participation and the number of breeding males needed for mating, bearing in mind the 3 levels of genetic correlation between UFW and FDM, within a system where most alpaca farmers employ subjective replacement selection criteria within their herds, with a limited exchange across herds owned by different farmers. This lack of control of genetic flow across herds and the small size of the herds, with an average 120 animals, could hinder effective selection within the farmers' herd. These selection schemes or strategies can be implemented under an proposed Improvement Plan, which aims to meet the selection objectives of quality and quantity improvement and with the aggregate genetic value selection criteria.

Although it has often been suggested that developing countries would benefit from the centralised nucleus selection scheme (Gizaw et al., 2009), this has not worked given the conditions in Huancavelica, since many programmes did work while there was financial and technical support, but, unfortunately, they succumbed without the involvement and commitment of the alpaca farmers, causing breeders to be excluded (Gizaw et al., 2010). Another alternative is the scheme based on dispersed nuclei, which was carried out by the PROCASUD (South American Camelid Improvement Programme) from the National University of Huancavelica (UNH, in Spanish), having selected 23 farmers with their respective herds. However, this programme has not achieved good results, mainly due to the lack of proper training of the farmers and their lack of willingness to keep productive and pedigree records, as well as deficient infrastructure and logistics.

Although this dispersed nuclei system fosters greater involvement in some breeder communities, which is proposed for ovine species by Gizaw et al. (2009) and in llamas by Wurzinger et al. (2008), it seems that, in our case, it might be a little too ambitious due to the shortage of technical experts, infrastructure and a greater willingness to exchange males across herds owned by different farmers.

Based on such considerations, perhaps a mixed system may be a good alternative, since it could have the advantages of the centralised nucleus and dispersed nucleus schemes. In this way, the issue of choosing farmers who may be more involved in keeping productive and pedigree records could be resolved, since, instead of needing 23 farmers, the scheme would need only 11 or 12, which is plausible in the population under study. In addition, the technical and productive performance could be perfectly complemented by the involvement of the UNH, where there are research groups able to conduct genetic evaluations.

In any case, in order for a genetic improvement programme to be guided in the right direction, as stated by Wurzinger et al. (2008), the farmers must be aware of the advantages and benefits of being a member of a genetic improvement organisation, since, otherwise, they would not be interested in becoming actively involved.

Moreover, all the selection schemes proposed must be implemented following an approach that incorporates aspects such as health care, shearing practices and fibre sales. These small incentives will encourage alpaca farmers to maintain their interest in the improvement plan, allowing for more time and money to be invested (Wurzinger et al., 2008). However, the most important and critical aspect of all the scenarios proposed – where the need to combine herds is included, mainly through the males, for the purpose of herd comparison (Mueller, 2008) – is the possibility of transmitting illnesses and parasites, which varies across strategies. Therefore, the involvement of farmers in performing regular health treatments will be essential, which will mean an additional cost, something that may be regarded as negative, yet mandatory (Wurzinger et al., 2008).

Evaluation of the selection schemes

Table 4 shows the expected genetic gains obtained for UFW (in g) and FDM (in µm) when using an aggregate genetic index as selection criteria on the different selection schemes proposed, and on three different levels of hypothetical genetic correlation between UFW and FDM (0.12, 0.25 and 0.50). When the selection is based on individual performance only, the genetic gain for UFW was negative while the genetic correlation between traits was assumed to be high (0.50); this is improved when the performance from relatives is also included. For the other assumed genetic correlation, genetic gain ranged from 29.54 to 146.81. However, when evaluating the genetic progress of FDM, it decreases in all scenarios, with various types of information and evaluation, as well as different levels of $r_{UFW,FDM}$, between a range of 0.68 and 1.45 µm, which is convenient for the farmers, the sector and consumers.

Overall, common nucleus schemes (Scenarios 3.1. and 3.2.) and the mixed scheme (centralised and common nuclei) show the lowest genetic and economic gains, when compared with the centralised nucleus scheme (Scenario 1.1, 1.2 and 1.3), which is concurrent with what was discussed for ovine specimens and llamas, by Lewis and Simm (2000), Kosgey et al. (2006), Kosgey and Okeyo (2007), Wurzinger et al. (2008) and Gizaw et al. (2010). With regard to Scenario 3.1., Scenario 1.3. increases the genetic gains for UFW and FDM as well as overall economic gains, by approximately 50% compared to Scenario 1.2, 30% over Scenario 1.1., 20% over Scenario 2.1. and 10% over Scenario 3.2. This pattern slightly differs when genetic correlation between traits is assumed to be high (0.50).

Selection based on the individual performance (IP), or IP plus BLUP evaluation (IP+BLUP), or IP+BLUP plus information of relatives (IP+BLUP+S), assuming a low genetic correlation (0.12) between traits, the total response was respectively 62%, 58% and 33% higher than when a high genetic correlation (0.50) was expected between them. When compared, the results between the medium (0.25) and the high (0.50) assumed genetic correlation, caused the total response to increase by 41%, 39%, and 23% in the last one.

Table 4. Estimated genetic gains per selection scheme scenario and 3 levels of genetic correlation between unwashed fleece weight (UFW) and fibre diameter (FDM).

Variable	Information[1]	Genetic correlation = 0.12					
		Scenario 1.1.[2]	Scenario 1.2.[2]	Scenario 1.3.[2]	Scenario 2.1.[2]	Scenario 3.1.[2]	Scenario 3.2.[2]
UFW (gr)	IP	55.93	60.31	67.14	50.10	45.84	50.48
	IP+B	64.03	68.95	78.33	57.47	52.68	57.90
	IP+B+S	121.33	130.75	146.81	118.82	99.68	109.63
FDM (µm)	IP	-0.91	-0.98	-1.09	-0.81	-0.74	-0.82
	IP+B	-0.93	-1.00	-1.14	-0.83	-0.76	-0.84
	IP+B+S	-1.2	-1.29	-1.45	-1.08	-0.09	-1.09
Total response (S/.)	IP	4.97	5.35	5.96	4.45	4.08	4.49
	IP+B	5.25	5.66	6.43	4.72	4.32	4.75
	IP+B+S	7.72	8.32	9.33	6.92	6.34	6.98
		Genetic correlation = 0.25					
UFW (gr)	IP	36.12	38.96	43.65	32.31	29.54	32.56
	IP+B	45.00	48.47	55.78	40.39	37.02	40.69
	IP+B+S	115.41	113.61	127.55	94.53	86.58	95.24
FDM (µm)	IP	-0.87	-0.93	-1.04	-0.78	-0.71	-0.78
	IP+B	-0.88	-0.95	-1.09	-0.79	-0.73	-0.80
	IP+B+S	-1.15	-1.4	-1.39	-1.03	-0.94	-1.04
Total response (S/.)	IP	4.33	4.66	5.22	3.87	3.55	3.91
	IP+B	4.61	4.97	5.71	4.14	3.79	4.17
	IP+B+S	7.11	7.67	8.60	6.38	5.84	6.43
		Genetic correlation = 0.50					
UFW (gr)	IP	-9.34	-10.00	-11.11	-8.46	-8.82	-8.52
	IP+B	1.28	1.37	1.52	0.14	1.05	1.15
	IP+B+S	67.83	73.14	82.11	60.79	75.63	61.25
FDM (µm)	IP	-0.82	-0.89	-1.00	-0.74	-0.68	-0.74
	IP+B	-0.82	-0.89	-0.98	-0.74	-0.68	-0.75
	IP+B+S	-1.38	-1.12	-1.25	-0.93	-0.85	-0.94
Total response (S/.)	IP	3.07	3.31	3.71	2.75	2.52	2.77
	IP+B	3.33	3.58	3.97	2.99	2.74	3.00
	IP+B+S	5.78	6.23	6.99	5.19	4.75	5.23

[1] Type of information: IP = only with information on the specimen's individual performance; IP + B = with IP and an evaluation is carried out through BLUP; IP + B + S = with IP + B plus related specimens' information.

[2] Scenarios proposed: 1.1: breeding specimens selection nucleus scheme (1 male/33 females); 1.2: breeding specimens selection nucleus scheme (1 male/45 females); 1.3: breeding specimens selection nucleus scheme (1 male/100 females); 2.1: mixed common selection scheme (inter- and intra-herd) and with reproductive specimens nucleus; 3.1: disperse common selection scheme (4% males per herd); 3.2: common selection scheme (3.38% males).

Our genetic progress estimates for UFW under different selection strategy scenarios, type of information and genetic correlation fall within the range reported on alpacas by León-Velarde and Guerrero (2001), Taipe *et al.* (2009) and Quispe *et al.* (2009a), who, based on simulation studies, estimate progress of 30, 28 and 49 gr/year, respectively; concerning llamas, Wurzinger *et al.* (2008) reports progress between 24 and 36 gr/year. However, our results are higher than those reported on alpacas by Ponzoni (2000), Quispe *et al.* (2007) and Quispe *et al.* The evaluation of the proposed selection schemes was conducted by considering three different sources of information: IP, selection using the individual performance, IP+BLUP, using the BLUP method with IP, and IP+BLUP+S, using the IP+BLUP method including information from relatives. Three different levels of genetic correlation between UFW and FDM (low: 0.12, medium: 0.25; and high: 0.50) were recorded. Performance of selection was studied in terms of genetic progress and inbreeding evolution, and was computed using the SelAction programme (Rutten and Bijma, 2001). Other genetic parameters used during this study, such as economic weights, phenotypic variances, phenotypic correlations, heritability and genetic correlations were those obtained by Quispe *et al.* (2009) or by Gutiérrez *et al.* (2009) as shown in Table 3. The estimated genetic progress was also lower than that reported by Mueller (2005) in ovine specimens, who estimates gains of 300 gr, although it should be noted that this study was carried out on Merino sheep. The genetic gains achieved for FDM (in μm) and total gains (in S/.) appear to be concurrent with what was reported on alpacas by Ponzoni (2000), León-Velarde and Guerrero (2001), Quispe *et al.* (2007a), Taipe *et al.* (2009) and Quispe *et al.* (2009a). It is also similar to the findings presented by Wurzinger *et al.* (2008) on llamas and Mueller (2005) on ovine wool.

If we compare the different sources of information used for genetic evaluation, we may observe that, with regard to selection based on IP, total gains increase by 55% with IP+B+S, whereas with IP+B, the increase is approximately 6%. Such increases become greater when the genetic correlations increase, becoming as high as 90% and 8%, respectively, improving IP+B+S and IP+B, when a 0.5 correlation is simulated. These results show the importance of pedigree and production information (Kosgey *et al.*, 2006; Gizaw *et al.*, 2009; Gizaw *et al.*, 2010), even more so when heritability in such characters ranges from low to medium, where the specimen's phenotype is a poor reflection of the specimen's genotype. The actual logistics and infrastructure used for this production system, especially in relation to pedigree registration, involves high costs and therefore the system remains under revision. Given the findings presented in this study the effort of improving it would be advantangeous.

The estimated genetic progress shows that alpacas from the population in Huancavelica have the potential to produce fibre, which can be improved by using dispersed common, mixed or single nucleus schemes. However, there is the need to determine the genetic parameters of this population, since this will allow for better simulations, propose better schemes and also view which information will be necessary in order to maximise applications.

Increase in the inbreeding rate under various selection schemes

The inbreeding increase in the scenarios proposed is displayed in Figure 2. It was lower than in Scenario 3.1, which showed 0.02%, although it was also low for scenarios 3.2 and 2.1, ranging from 0.05% and 0.06%. The highest rate was seen in scenario 1.3, while regular increases were obtained for scenarios 1.1 and 1.2. It was also found that the different levels of genetic $r_{UFW,FDM}$ (low = 0.12, medium = 0.25 and high = 0.50) are not related to the inbreeding increase, since on the three levels employed (low, medium and high), the inbreeding in the different scenarios almost did not vary at all.

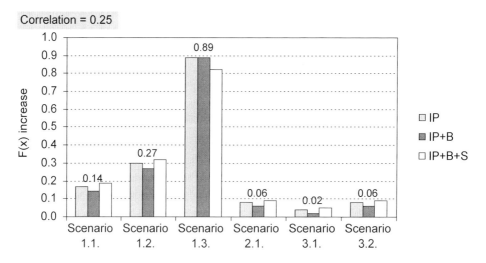

Figure 2. Inbreeding percentage increase within 6 selection schemes and 3 genetic correlation levels between UFW and FDM, using various sources of information and genetic evaluation.

With regard to Scenario 3.1: in Scenario 1.3., inbreeding showed a 21-fold increase; in Scenario 1.2, an 8-fold increase; in Scenario 1.1., a 4-fold increase; and in Scenario 2.1. and 3.2, a 2-fold increase, where the last two have the same inbreeding increase rate. The highest inbreeding rates can be seen in scenarios where the highest number of females per breeding male is presented, resulting in the largest rate (nearing 1%) in the scheme (Scenario 1.3.) where artificial insemination is used in such a way that for 100 females only 1 male would be needed. However, the advantage is that this way, further genetic progress can be achieved, which can be sustained by renewing males coming from other alpaca populations available in Peru (Puno, Cuzco and Arequipa, among others).

The inbreeding increases found are rather similar to those reported on llamas by Wurzinger *et al.* (2008) and on sheep by Lewis and Simm (2000), and Gizaw *et al.*, 2010). In any case, the inbreeding rate increases are low (less than 1%), which is advisable in order to avoid endogamic depression (Lewis and Simm, 2000).

Conclusion

In terms of the breeding system to be used in the Huancavelica Region, the mixed scheme could be a good alternative, thus encouraging the farmers' involvement and the complementary interaction between technical and government experts. Regarding genetic progress for UFW and FDM, the use of information from performance of relatives seems to considerably improve the low response obtained when only the genetic value of the animal is used to carry out the selection process. Inbreeding increases are low (less than 1%), which could prevent endogamic depression.

Aknowledgements

We sincerely value the financial support of the Government of Navarre and the Public University of Navarre (Spain), through the Project: 'Improvement of the economic income and conditions of life of poor families by means of fortification of capacities and improvement of Huacaya alpacas in the high-Andean zone of Huancavelica (Peru)'.

References

Brenes E.R., K. Madrigal, F. Pérez and K. Valladares, 2001. El Cluster de los Camélidos en Perú: Diagnóstico Competitivo y Recomendaciones Estratégicas. Instituto Centroamericano de Available at: http://www.caf.com/attach/4/default/CamelidosPeru.pdf. Accessed 21 March 2009.

Gizaw S., H. Komen and J.A.M van Arenkonk, 2009. Optimal village breeding schemes under smallholder sheep farming systems. Livestock Science, 124: 82-88.

Gizaw S., H. Komen and J.A.M. van Arendonk, 2010. Participatory definition of breeding objectives and selection indexes for sheep breeding in tradicional systems. Livestock Science, 168: 67-74.

Gutiérrez J.P., F. Goyache, A. Burgos and I. Cervantes, 2009. Genetic analysis of six production traits in Peruvian alpacas. Livestock Science, 123: 193-197.

Kosgey I.S., R.L. Baker, H.M.J. Udo and J.A.M. van Arendonk, 2006. Succeses and failures of small ruminant breeding programmes in the tropics: a review. Small Ruminant Research, 61: 13-28.

Kosgey I.S. and Okeyo, 2007 Genetic improvement of small ruminants in low-input, smallholder production systems: Technical and infrastructural issues. Small Ruminant Research, 70: 76-88.

Lewis R.M. and G. Simm, 2000. Selection strategies in sire referencing schemes in sheep. Livestock Production Science. 67: 129-141.

León-Velarde C.U and J. Guerrero, 2001. Improving quantity and quality of Alpaca fiber; using simulation model for breeding strategies. Available at: http://inrm.cip.cgiar.org/home/publicat/01cpb023.pdf. Accessed 18 April 2010.

Mason I.L. and V. Buvanembra, 1982. Breeding plans for ruminant livestock in the tropics. FAO Animal Production and Health Paper (FAO), no. 34. Rome, Italy, 98 p.

Mueller J.P., 2005. Respuestas a la selección en Merino con diferentes procedimientos. Proc. XXXIV Congreso Argentino de Genética. Comunicación Técnica. Bariloche Nro. PA 473.

Mueller J.P., 2008. Estrategias para el mejoramiento de camélidos sudamericano. En: Actualidades sobre adaptación, producción, reproducción y mejora genética en camélidos. Universidad Nacional de Huancavelica. First edition, Huancavelica, Perú, pp. 93-112.

Nieto L. and I. Alejos, 1999. Estado económico y productivo del Centro de Producción e Investigación de Camélidos Sudamericanos – Lachocc. XXI Reunión Científica Anual APPA, Puno, Peru.

Ponzoni R.W. 2000. Genetic improvement of Australian Alpacas: present state and potential developments. Proceedings Australian Alpaca Association, 1: 71-96.

Quispe E.C. 2005. Mejoramiento Genético y Medioambiental de Alpacas en la Región de Huancavelica. Proyecto de Inversión Pública a nivel de Perfil. Universidad Nacional de Huancavelica, Huancavelica, Peru.

Quispe E.C., L. Alfonso, R. Paurcar and H. Guillén, 2007. Predicción de respuesta a la selección de alpacas Huacaya en la Región Altoandina de Huancavelica.progreso genetico. Proc. de I Congreso Nacional de reproducción y Mejoramiento Genético de Camélidos Sudamericanos, Huancavelica, Peru.

Quispe E.C., L. Alfonso, A. Flores, H. Guillén and Y. Ramos, 2009a. Bases to an improvement program of the alpacas in highland region at Huancavelica-Perú. Archivos de Zootecnia, 58 (224): 705-716.

Quispe E.C., T.C. Rodríguez, L.R. Iñiguez and J.P. Mueller, 2009b. Producción de fibra de alpaca, llama, vicuña y guanaco en Sudamérica. Animal Genetic Resources Information, 45: 1-14.

Rutten M.J.M. and P. Bijma, 2001. SelAction manual. Wageningen University, Wageningen, the Netherlands, p. 34.

Taipe H., B. Cárdenas and E.C. Quispe, 2009. Predicción de progreso genético para peso de vellon sucio y media de diámetro de fibra en alpacas de la raza Huacaya de color blanco en la región de Huancavelica. Thesis, Universidad Nacional de Huancavelica, Huancavelica, Peru, p. 88.

Wurzinger M., A. Willam, J. Delgado, M. Nürnberg, A. Valle Zárate A., A. Stemmer, G. Ugarte and J. Sölkner, 2008. Design of a village breeding programme for a llama population in the High Andes of Bolivia. Journal of Animal Breding and Genetics, 125: 311-319.

Mitochondrial DNA (mtDNA) genetic diversity of *Vicugna vicugna mensalis* in Bolivia

J. Barreta[1,2], V. Iñiguez[2], R.J. Sarno[3], B. Gutiérrez-Gil[1] and J.J. Arranz[1]
[1]Dpto. Produccion Animal, Universidad de Leon, Spain; jbarretapinto@yahoo.com
[2]Instituto de Biologia Molecular y Biotecnologia, UMSA, La Paz, Bolivia
[3]Department of Biology, Hofstra University Hempstead, NY, USA

Abstract

The vicuña (*Vicugna vicugna*) is one of two wild South American camelids. Nowadays in Bolivia, there are over 62,869 animals representing 20% of the world's population breeding in the wild as the only management system. Two subspecies of vicuña are currently accepted, based largely on genetic and size differences: *Vicugna vicugna vicugna*, distributed in Argentina, Chile and South of Bolivia, and *Vicugna vicugna mensalis*, disseminated in Peru and North of Bolivia. There are few studies analysing the molecular diversity in Bolivian vicuñas and the aim of this work is to better understand the genetic variability of *V. v. mensalis* distributed in this country. For this purpose, and based on the *Vicugna v. mensalis* published sequence, we analysed the complete cytochrome-b (Cyt-b) (1140 bp) and partial (D-loop) (511 bp) regions of mitochondrial DNA across 35 animals from four different locations within a protected area in the north of Bolivia (Apolobamba). A total of 604 polymorphic sites were found in the Cyt-b gene yielding nine different haplotypes. In the case of the D-loop region, nineteen polymorphic sites were detected and grouped into three haplotypes different. Haplotype diversity (*h*) reached 0.877 in the Cyt-b gene and 0.211 in the D-loop region. Nucleotide diversity (π) ranged from 0.2572, in the Cyt-b, to 0.00249, in the D-loop. These results show the need for monitoring programmes in wild populations recovered by massive expansion of the species and the constant evaluation of genetic variation to support conservation and management programmes.

Keywords: camelids, vicuña, mitochondrial DNA, genetic diversity

Introduction

The vicuña (*Vicugna vicugna*) is the wild South American camelid which is best adapted to high altitude ecosystems (Bonacic *et al.*, 2002). It has the finest wool in the world (approximately 12 μm in diameter) and there is great interest and demand for its use and exploitation (Kadwell *et al.*, 2001).

In the sixteenth century, the total estimated population of vicuña reached 1.5 to 2 million individuals. However, in the 1960s and after over-exploitation, the vicuña was declared an endangered species with the number of individuals between 7,000 and 12,000 across its entire range of distribution. After this period, the species was protected in countries with natural populations (Convention for the Conservation and Management of Vicuña originally signed between Peru and Bolivia in 1969, later joined by Argentina, Chile and Ecuador) (Bonacic *et al.*, 2002).

The current geographical distribution ranges from latitudes South 9°30' to 29°00' and it is found only above 3,500 metres in altitude on the Peruvian, Bolivian, Chilean and Argentinean treeless grassland steppes called Puna and Altiplano (Renaudeau, 2002).

The first protected areas were the Pampa Galeras Reserve in Peru and the Andean Fauna Reserve Ulla Ulla in Bolivia, which were later joined by other reserves and national parks. Today, vicuña management is carried out in various forms. At one extreme there is the management in captivity in Argentina. At the other extreme lies the community-based management of vicuñas in the wild such as in Bolivia, which is the sole country to declare that all its vicuñas remain as wild populations. Peru and Chile's policies are situated in the middle with a strong trend towards shifting from community-based management of vicuñas in the wild to semi-captivity (Renaudeau, 2002).

Over a period of 40 years, the vicuña was recovered in great numbers, now estimated to be approximately 306,680 across the highlands and central highlands. Nowadays, the vicuña population has approximately 2,683 individuals in Ecuador, 16,351 in Chile and around 50,000 animals in Argentina. The list is followed by Bolivia with a population of 62,869 and finally Peru with 174,377 individuals (CITES, 2007).

The recovery of these populations led to the 1987 Convention on International Trade in Endangered Species of Fauna and Flora (CITES), which approved the transfer of some populations within Peru and Chile allowing international trade in wool sheared from live vicuñas. In the following years, this regulation was extended to some other populations of vicuña in Argentina and Bolivia (Kadwell *et al.*, 2001).

Two geographic subspecies of the vicuña are recognised: the northern *Vicugna vicugna mensalis*, and the southern *Vicugna vicugna vicugna* (Torres, 1992). The approximate dividing line between these two races is 18° South latitude. *V. v. vicugna* (Molina, 1872) is found in Bolivia, Argentina and some regions of Chile. This southern subspecies is both larger and lighter in colour than the northern type. *V. v. mensalis* (Thomas, 1917) occurs in Bolivia, Peru and Chile. This subspecies is darker in colour when compared with the southern vicuña and possesses a white pectoral tuft that is absent in *Vicugna v. vicugna* individuals (Yacobaccio, 2006).

Based on the study of nuclear genetic markers, Wheeler *et al.* (2001) reported low levels of genetic diversity within populations and high levels between populations in wild populations of wild Peruvian vicuña (*V.v. mensalis*). This is a typical situation for species that have suffered a bottleneck in population size and a consequent risk of extinction in the past (Vilá, 2002). Sarno *et al.* (2004) described low to moderate genetic diversity in Bolivian and Chilean vicuñas, with the northern Bolivian population being the least variable compared with the central and southern Bolivian populations (intermediate levels) and the Chilean population (the most variable). A more extensive study, including populations from Peru, Chile, and Argentina reported a high degree of mitochondrial diversity at the haplotype level and low nucleotide diversity in the species as a whole and within groups. When comparing the two subspecies, the southern populations were found to be more diverse than those in the north (Marín *et al.*, 2007a), supporting the results described by Sarno *et al.* (2004).

Today, our knowledge on the genetic variability of Bolivian vicuñas is limited to the work of Sarno *et al.* (2004), where a limited number of animals from south, north and central Bolivia were analysed. Although the geographical boundaries for the two vicuña subspecies in Bolivia are unclear, these authors suggest that *V. v. mensalis* would be found distributed mainly in ANMIN-Apolobamba and *V. v. vicugna* in South Lípez. Taking into account the lower level of genetic diversity reported for the northern vicuñas (*V. v. mensalis*) compared with the southern populations (*V. v. vicugna*) (Sarno *et al.*, 2004; Marín *et al.*, 2007a) and the rapid demographic expansion suggested for *V. v. mensalis* (Marín *et al.*, 2007b) we have studied the genetic diversity of mitochondrial DNA (cytochrome-b, Cyt-b, and D-loop regions) in a vicuña population localised in ANMIN-Apolobamba, in northern Bolivia. Hence, the aim of this study was to determine whether the degree of genetic diversity for this stock, which is likely to belong to the *V. v. mensalis* group, has achieved a significant recovery as would be expected based on the conservation measures implemented since 1969 (Convention for the Conservation and Management of Vicuña).

Materials and methods

Sample collection and DNA extraction

The study included a total of 35 *Vicugna v. mensalis* individuals from ANMIN-Apolobamba (northern region of Bolivia 18° 55 'S, 69° 4'60 W). This reserve has an estimated population of 8,299 vicuñas in an area of 100,000 Ha (DNCB, 1996), identifying four areas of herding, trapping and shearing: Cañuhuma, Hichucollo, Huacuchani and Ucha Ucha. Adult animals were sampled at random from these four locations (Cañuhuma n=11; Hichucollo n=8; Huacuchani n=10; and Ucha Ucha n=6). Ear tissue samples were collected from the animals when they were sheared and kept in 'Easy Blood' buffer (Munson, 2000). Genomic DNA isolation was performed from the ear tissue samples (2-3 mm) with the commercial DNeasy Tissue kit (QIAGEN) following the manufacturer's instructions. A guanaco (*Lama guanicoe*) was used as reference for the analysis of the Cyt-b gene, whereas a llama (*Lama glama*) was the reference sample in the analysis of the D-loop region.

Polymerase chain reaction amplification

The mitochondrial Cyt-b ($\approx$ 1,140 bp) and the D-loop region ($\approx$ 511 bp) were amplified by PCR using the primer pairs detailed in Table 1, following Marín *et al.* (2007b).

The amplification reaction was performed in a total volume of 30 μl containing 30ng of DNA, 1X magnesium free PCR buffer (Applied Biosystems (AB), Foster City. CA), 2.0 mM $MgCl_2$, 0.2 mM dNTPs, 0.5 mM of each primer and 1.25 U of AmpliTaq Gold (AB). The PCR programme

Table 1. Oligonucleotide sequences used for the amplification and sequencing the mtDNA regions studied in this work (cytochrome-b gene and D-loop region).

Name of primer	Sequence nucleotide	mtDNA region
LGlu ARTIO[1]	5' TCTAACCACGACTAATGACAT 3'	Cyt-b
Hthr ARTIO[1]	5' TCCTTTTTCGGCTTACAAGACC 3'	Cyt-b
Lthr ARTIO[1]	5' GGTCTTGTAAGCCGAAAAAGGA 3'	D-loop
HLOOP550G[1]	5' ATGGACTGAATAGCACCTTATG 3'	D-loop
L400[2]	5' GGGCTATGTACTCCCATGAGG 3'	Cyt-b
LBE-02[2]	5' CTCCGTAGATAAAGCCACCC 3'	Cyt-b
CytB337[2]	5' TTCAAGTTTCTAGGAAGGGCG 3'	Cyt-b
Lloop0007G[2]	5' GTACTAAAAGAAAATATCATGTC 3'	D-loop
H362[2]	5' GGTTTCACGCGGCATGGTGATT 3'	D-loop
HI5998[2]	5' CCAGCTTCAATTGATTTGACTGCG 3'	D-loop
Dloop747F[2]	5' TAAAATCGCCCACACACTTTCC 3'	D-loop

[1] Primers used in the PCR fragment amplification and later for sequencing.
[2] Internal primers used for sequencing.

consisted of an initial 5-min denaturation step at 95 °C, followed by 35 cycles of 45 s at 95 °C, 30 s at 60 °C and 45 s at 72 °C, followed by a final elongation cycle of 5 min at 72 °C.

Sequence analysis

The amplified fragments were purified with 4U of ExoSAPIT (Amersham Biosciences, GE Healthcare) by incubating at 37 °C for 30 min, followed by an enzyme deactivation phase for 15 min at 80 °C. Dideoxy sequencing in both directions was performed with the BigDye Terminator v3.1 Cycle Sequencing Kit (Applied Biosystems, Foster City, CA) using the same primers used for fragment amplification. Additional internal primers, which are also given in Table 1, were designed based on the reference sequences for *V. v. mensalis* obtained from GenBank (GenBank Acc N° AY535255 for Cyt-b, and GenBank Acc N° AY856270 for D-loop region). The sequencing reactions included an initial denaturation step of 96 °C for 5 min and 35 cycles of 96 °C for 20 s, 59 °C for 10 s, and 60 °C for 4 min. The products of the sequencing reaction were purified using CleanSEQ reagent set (Agencourt Bioscience, Beckman Coulter) and analysed using an ABI3130 DNA Sequencer (Applied Biosystems Foster City CA).

Data analyses

Using SeqScape v2.5 software (Applied Biosystems, Foster City, CA) the sequences obtained for the 35 animals analysed were aligned and compared with the published sequences (GenBank Acc N° AY535255 and AY856270 for Cyt-b and D-loop regions, respectively) to search for any polymorphisms in the studied fragments. Sequences were aligned using the DNA alignment software (Fluxus technology, 2003-2010) and were visually confirmed. The number of polymorphic sites (S), the average number of differences between pairs of sequences (P), haplotype diversity (h) and nucleotide diversity (π) were estimated with DnaSP 4.0 (Rozas, 2009). We generated a median joining network for each of the two studied regions using NETWORK (4.5.1.6) software (Bandelt *et al.*, 1999) in order to display the relationships among the identified haplotypes.

Results

We obtained data for the complete mitochondrial Cyt-b gene (1,140 bp) and partial D-loop region (511 bp) from 29 and 35 individuals, respectively, which belonged to the four localities from the ANMIN-Apolobamba reserve considered in the present study. The diversity parameters estimated for the two regions are given in Table 2. Briefly, alignment of the sequences showed 606 (52.63%) polymorphic sites at the Cyt-b gene and 19 (3.72%) polymorphic sites at the D-loop region. These polymorphic sites were grouped into 9 and 3 different haplotypes in the Cyt-b and D-loop regions, respectively. Haplotype diversity (h) was much lower at the D-loop region (0.2110±0.0890) than in the Cyt-b gene (0.8770±0.0340) as was the nucleotide diversity (π) (0.0025±0.0016 for the D-loop region and 0.2572±0.0127 for the Cyt-b gene).

Table 2. Haplotype (gene) diversity (h) and nucleotide diversity (π) estimated for the mtDNA regions studied in this work.

	Fragment size (bp)	Haplotypes	h ± std dev	π ± std dev
Cyt-b	1,140	9	0.8770±0.0340	0.2572±0.0127
D-loop	511	3	0.2110±0.0890	0.0025±0.0016

The alignment of polymorphic sites identified in the Cyt-b and D-loop regions is shown in Figure 1, where the frequency observed for each haplotype and the localities corresponding to each identified haplotype are also indicated. In the Cyt-b gene (Figure 1a) the 604 segregating sites were distributed in nine different haplotypes. The most frequent haplotype was H1 (*f*=13/29), which was found in Cañuhuma, followed by H3 (*f*=8/13) identified in the same locality. For the D-loop region (Figure 1b), there was a dominant haplotype, H1 (*f*=32/35), which was observed in all localities whereas two other haplotypes, H2 and H3, were only found in one of the four localities studied at very low frequency (*f*=2/35 and *f*=1/35, respectively).

For the Cyt-b gene, the median joining network constructed shows the genealogical relationships among the 9 haplotypes connected through a maximum of 19 mutational steps (Figure 1a). The median joining network obtained for the D-loop sequences establishes the relationship among 3 haplotypes connected through a maximum of 5 mutational steps (Figure 2).

Discussion

Andean animal populations live in one of the most extreme and variable topographic environments in the world. Hence, ecological factors, together with historic factors may have influenced the patterns of variation of *mt*DNA shaping the genetic relationships between the current populations. We have analysed here *mt*DNA sequences from a vicuña population localised in the ANMIN-Apolobamba reserve. Because of their geographical location, these animals are expected to belong to the *V.v. mensalis* group of vicuñas, which shows lower genetic diversity levels than the *V. v. vicugna* group according to previous studies (Sarno *et al.*, 2004; Marín *et al.*, 2007a).

Overall, our analyses based on *mt*DNA sequences showed appreciable genetic variability segregating in the vicuña population analysed. This variation was much higher for the Cyt-b gene (h = 0.8770; π = 0.2572) than for the D-loop region (h = 0.2110; π = 0.00249), which is surprising taking into account that in most of the works studying *mt*DNA variability levels across different mammalian genera, the D-loop region is considered a hypervariable region (Pedrosa

A

```
Haplotypes 23459124667911112222333333333344444444455555555566666666677777777788888888889999999999111111111111      n       Localities
                87509260257144914556999999045678890224667890235789911244789901344577780011456899000000000111  Total = 29
                      8368903452489012369843516981783129519194285934015165819709836930756576246333445566812 2
                                                                                                  2581439581329
   Ref.    TGACCAGCCTCTGAAAGTCATCTCGCTATACACACTGGAAACTCAGAAACATCTCATTCATCTACCTTTGTAACACTCTGCACCAACGCCTGCTTCTC
     H1.   ..................G.T..TT.......G......CAC....TG......G...G.T...T..C..G......C..G..GG........C.C.     13       C
     H2.   ..................G.T..TTAT.....G......CAC....TG......G...G.T...T..C..G......C..G..GG........C.C.      1       C
     H3.   ..................G.T..TT......TG......CAC....TG......G...G.T...T..C..G......C..G..GG........C.C.      8       C
     H4.   ..................G.T..TT.......G......CAC....TG......G...G.T...T..C..G......C..G..GG........C.CT      1       H
     H5.   ..................G.TG.TT.......G......CAC....TG......G...G.T...T..C..G......C..G..GG........C.C.      2       H
     H6.   ..................G.T..T...CCT..G......CAC....TG......G...G.T...T..C..G......C..G..GG........C.C.      1       I
     H7.   ..................G.TG.TT......TG......CAC....TG......G...G.T...T..C..G......C..G..GG........C.C.      1       I
     H8.   ....TG.T...C......G.T..TT......TG......CAC....TG......G...G.T...T..C..G......C..G..GG........C.C.      1       I
     H9.   ATGA..............G.T..TT.......G......CAC....TG......G...G.T...T..C..G......C..G..GG........C.C.      1       U
```

B D-loop

```
Haplotypes      6911122333345555666667112222233344          n          Localities
                  569890378448890478929911233478         Total = 35
                                      1517907709
   Ref.         GGGAGCCATACTGTCTCGATGACTGTCTCTT
   H1.          .AAGATT.......T.TAGC.GTC...C...          32          C, H, I, U
   H2.          .AAGATT.......T.TAGC.GTC.C.CT..           2          C
   H3.          AAAGATT...T.A.T.TAGC.GTC...C...           1          U
```

Figure 1. Alignment of polymorphic sites identified across the Cyt-b (a) and D-loop (b) regions analysed in this study. Nucleotide umbering is based on GenBank reference sequences (GenBank Acc N° AY535255 for Cyt-b gene and GenBank Acc N° AY856270 for D-loop region). Localities code C = Cañuhuma, H = Huacuchani, I = Huchicollo and U = Ucha Ucha.

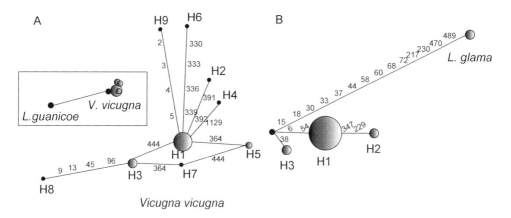

Figure 2. Median joining network for the analysed vicuña population based on the Cyt-b (a) and D-loop (b) haplotypes identified in the present study. The size of nodes is approximately proportional to their frequency.

et al., 2006; Alter and Palumbi, 2009, Galtier *et al.*, 2009). Interestingly, Marín *et al.* (2008) found a slightly higher ($h = 0.956$) variability for the Cyt-b gene than for the D-loop region ($h = 0.890$) in guanaco (*Lama guanicoe*), the other species of wild South American camelids. Further studies should clarify whether this observation occurs in other South American camelids or, alternatively, if it is due to some bias not considered in our analyses.

The haplotype alignment of Cyt-b sequences showed two dominant haplotypes (H1 and H3), with only one mutational step between them, from a total of nine. In the D-loop sequence we found a single dominant haplotype (H1) from a total of three. There was also a single mutational step occurring between the dominant D-loop haplotype and the two other haplotypes identified in our resource population. These observations probably indicate that the vicuña populations found in ANMIN-Apolobamba are being subjected to a rapid demographic expansion. In addition, the presence of a large dominant haplotype possibly represents an ancestral lineage, since older haplotypes are expected to have a wider geographic range and frequency (Templeton *et al.*, 1995).

Sarno *et al.* (2004) found variable levels of genetic diversity for vicuñas sampled across northern, central and southern Bolivia, considering at the same time three combined *mt*DNA segments involving Cyt-b gene, D-loop region and 16Ss. The northern vicuñas analysed in this study, which are also from the ANMIN-Apolobamba reserve, showed null haplotype and nucleotide diversity, whereas central and southern vicuñas showed increasing levels of genetic diversity. In general, we have found higher levels of genetic diversity across the vicuñas localised in ANMIN-Apolobamba, although a direct comparison of results is not easy. The observed differences are likely due to the larger sample population analysed in the current study. In addition, the inherent differences in genetic variability across the studied regions may influence the results. In this regard, we observed a large variation of the genetic variability across the complete sequence of the Cyt-B gene (1140 pb), with low variability in the first third of the gene ($\pi = 0.0017$, considering 1-325 pb) when compared with the second half of the gene ($\pi = 0.3712$, considering 326-1140 pb), whereas Sarno *et al.* (2004) do not detail the position of the 316 pb that they sequenced.

On the other hand, when Sarno *et al.* (2004) present the joint analysis of northern and central vicuñas, within the *V.v. mensalis* group, the gene diversity parameters observed ($h = 0.395$; $\pi = 0.002$) show certain levels of genetic diversity and, similar to those observed for D-loop in this paper ($h = 0.2110$; $\pi = 0.0025$), whereas the *V. v. vicugna* group showed a slightly higher variability ($h = 0.524$; $\pi = 0.008$), although still lower than that observed in the resource population analysed here ($h = 0.8770$; $\pi = 0.2572$). The joint analysis of the Cyt-b, D-loop and s16 regions presented by Sarno *et al.* (2004) can also make it difficult to compare these two studies.

The genetic variability of the D-loop region reported in *V.v. mensalis* vicuñas ($h=0.7350$; $\pi=0.0082$) from Perú and Chile (Marin *et al.*, 2007) is higher than that reported here for the same genomic region ($h=0.2110$; $\pi=0.0025$). In this case, it seems that the larger number of individuals (n=261) analysed may underlie the higher genetic diversity observed by Marín *et al.* (2007a). Hence, to increase our knowledge of the genetic diversity of Bolivian vicuñas and compare them with other populations, further research studies should be focused on larger population sizes involving, if possible, both geographical vicuñas groups, *V. v. mensalis* and *V. v. vicugna*. This could help to confirm a clear genetic differentiation between the two subspecies, and the lower variability levels reported for *V.v. mensalis* in all regions, countries and management units studied (Sarno *et al.*, 2004; Marín *et al.*, 2007a). Moreover, analysis of nuclear microsatellite loci could increase the resolution of the genetic differentiation that exists between these two subspecies.

Conclusion

The levels of genetic diversity achieved in the vicuña populations living in the ANMIN-Apolobamba reserve are significant, considering the remarkable increase in the population of this species in this region during the last years (from 97 individuals in 1965 to 8,000 at present). As a concluding remark, and based on the preliminary results presented here, it seems that the analysis of the Cyt-b gene, especially the last region of its sequence, could be a more useful tool to assess the genetic variability of neighbouring vicuña populations than the analysis of the D-loop region.

Acknowledgements

We gratefully acknowledge the financial support received from the AECID by projects A/010497/07 and A/017114/08. Julia Barreta's scholarsip is provided by MAEC-AECID.

References

Alter, S.E. and S.R. Palumbi, 2009. Comparing evolutionary patterns and variability in the mitochondrial Control Region and Cythocrome b in three species of Baleen Whales. Journal of Molecular Evolution, 68: 97-111.

Bandelt H.J., P. Forster and A. Röhl, 1999. Median-joining networks for inferring intraspecific phylogenies. Molecular Biology and Evolution, 16: 37-48.

Bonacic C., D. Macdonald, J. Galaz and R. Sibly, 2002. Density dependence in the camelid Vicugna vicugna: the recovery of a protected population in Chile. Oryx, 36: 118-125.

Fluxus technology, 2003-2010. DNA Alignment Software. Available at: http://www.fluxus-engineering.com.

Galtier N., B. Nabholz, S. Glémin and G.D.D. Hurst, 2009. Mitochondrial DNA as a marker of molecular diversity: a reappraisal. Molecular Ecology, 18: 4541-4550.

CITES (Convention on International Trade in Endagered Species), 2007. Proposals for amendment of Appendices I and II of CITES convention. Available at: http://www.cites.org/eng/cop/14/prop/E14-P08.pdf.

Kadwell M., M. Fernández, H.F. Stanley, R. Baldi, J.C. Wheeler, R. Rosadio and M.W. Bruford, 2001. Genetic analysis reveals the wild ancestors of the llama and alpaca. Proceedings of the Royal Society of London B, 268: 2575-2584.

Marin J.C., C.S. Casey, M. Kadwell, K. Yaya, D. Hoces, J. Olazabal, R. Rosario, J. Rodriguez, A. Spotorno, M.W. Bruford and J.C. Wheeler, 2007a. Mitochondrial phylogeography and demographic history of the vicuña: implications for conservation. Heredity, 99: 70-80.

Marín J.C., B. Zapata, B.A. González, C. Bonacic, J.C. Wheeler, C. Casey, M.W. Bruford, R.E. Palma, E. Poulin, M.A. Alliende and Á.E. Spotorno, 2007b. Sistemática, taxonomía y domesticación de alpacas y llamas: nueva evidencia cromosómica y molecular. Revista Chilena de Historia Natural, 80: 121-140.

Marín J.C., A.E. Spotorno, B.A. González, C. Bonacic, J.C. Wheeler, C.S. Casey, M.W. Bruford, R.E. Palma and E. Poulin, 2008. Mitochondrial DNA variation and systematics of the guanaco (*Lama guanicoe*, Artiodactyla: Camelidae). Journal of Mammalogy, 89: 269-281.

Molina G.I., 1872. Saggio Sulla Storia Naturale del Chili. Tommaso d'Aquino, Bologna, Italy.

Munson L., 2000. Necropsy procedures for wild animals. With input from: W.B. Karesh, M.F. McEntee, L.J. Lowenstine, M.E. Roelke-Parker, E. Williams and M.H. Woodford (Illustrations by D. Haines). In: L. White and A. Edwards (eds.). Conservation Research in the African rain forests: a technical handbook. Wildlife Conservation Society, New York, NY, USA, pp. 203-224.

Pedrosa S., J.J. Arranz, N. Brito, A. Molina, F. San Primitivo and Y. Bayón, 2006. Mitochondrial diversity and the origin of Iberian sheep. Genetics Selection Evolution, 39: 91-103.

Renaudeau d'Arc, N., 2002. 'Community Management of Vicuña in the Bolivian Altiplano', Procedural Paper presented in March as requirement from MPhil to PhD, School of Development Studies, University of East Anglia, Norwich, UK.

Rozas J., 2009. DNA Sequence Polymorphism Analysis using DnaSP. In: D. Posada (ed.) Bioinformatics for DNA Sequence Analysis; Methods. Molecular Biology Series Vol. 537. Humana Press, Totowa, NJ, USA, pp. 337-350.

Sarno R.J., L. Villalba, C. Bonacic, B. Gonzalez, B. Zapata, D.W. Mac Donald, S.J. O'Brien and W.E. Johnson, 2004. Phylogeography and subspecies assessment of vicuñas in Chile and Bolivia utilizing mtDNA and microsatellite markers: implications for vicuña conservation and management. Conservation Genetics, 5: 89-102.

Templeton A.R., E. Routman and C.A. Phillips, 1995. Separating population structure from population history: a cladistic Analysis of the geographical distribution of mitochondrial DNA haplotypes in the tiger salamander, Ambystoma tigrinum. Genetics, 140: 767-782.

Thomas O., 1917. Preliminary diagnosis of new mammals obtained by the Yale National Society Peruvian Expedition. Smithsonian Misc C 68: 1-3.

Torres H., 1992. Camélidos Silvestres Sudamericanos. Un plan de Accion para su Conservacion. IUCN/CSE South American Camelid Specialist Group, 58 p.

Vilá B.L., 2002. La silvestría de las vicuñas: Una característica esencial para su conservación y manejo. Ecología Austral, 12: 79 82.

DNCB (Dirección Nacional de Conservación de la Biodiversidad), 1996. Censo Nacional de la vicuña en Bolivia. Informe publicado por el Ministerio de Desarrollo Sostenible y Medio Ambiente. La Paz. 55 p.

Yacobaccio H.L., 2006. Variables morfométricas de vicuñas (*Vicugna vicugna vicugna*) en Cieneguillas. Jujuy. In: B. Vilá (ed.), Investigación conservación y manejo de vicuñas. Proyecto MACS. Buenos Aires, Argentina, pp. 101-112.

Wheeler J.C., M. Fernández, R. Rosadio, D. Hoces, M. Kadwell and M.W. Bruford, 2001. Diversidad Genética y manejo de poblaciones de vicuñas en el Perú. RIVEP Revista de Investigaciones Veterinarias del Perú, Suplemento 1: 170-183.

Association of myostatin gene (*MSTN*) polymorphism with economic traits in rabbits

K.A. Bindu[1], Arun Raveendran[1], Siby Antony[2] and K.V. Raghunandanan[1]
[1]Department of Animal Breeding and Genetics, College of Veterinary and Animal Sciences, Mannuthy, Trichur, Kerala, India; binsib@rediffmail.com
[2]Resident Veterinarian, I.F.C.D.W., Agadir, Morocco

Abstract

Myostatin (*MSTN*) also known as GDF_8 is a negative regulation factor which determines the maximal amount of body mass typical for every species. *MSTN* is considered as an important candidate gene for meat production. Fontanesi *et al*. (2008) detected a single nucleotide polymorphism (C-T transition) at position 34 of intron 2 of myostatin gene. If there is a mutation in *MSTN* gene, its negative regulation function is disrupted leading to double muscle phenotype. The rabbit *MSTN* gene sequence comprises three exons and two introns as observed in other species. A total of 60 DNA samples of New Zealand White, Soviet Chinchilla and crossbred rabbits were subjected to PCR-based RFLP. An 80 bp fragment of *MSTN* gene was amplified using PCR and digested with *Alu 1* enzyme and polymorphism was detected in the population studied, revealing three genotypes, viz CC, CT and TT. In the present study, CT genotypes were associated with higher body weight but the difference in body weight among different genetic groups was not statistically significant.

Keywords: myostatin, polymorphism, MSTN

Introduction

There has been a rise in global awareness about the virtues of rabbit meat and fibre. Rabbits have traits of economic importance like high rate of reproduction, early maturity, rapid growth rate and efficient feed utilisation. Moreover, they pose limited competition for human food and produce high-quality nutritious meat. This potential of rabbits can be well utilised at a commercial or even small-scale level to address the problems of food scarcity. Only limited studies have been conducted in rabbits of tropical countries, where climate, diet, management and stock resources can differ markedly from those in temperate countries. A candidate approach has already been successfully applied to identify several DNA markers associated with production traits in livestock (Rothschild and Soller, 1997).

Myostatin gene can be considered as one of the candidate genes for meat production traits in rabbits. The principle is based on the fact that the variability within genes coding for protein products involved in key physiological mechanisms and metabolic pathways directly or indirectly involved in determining an economic trait might explain a fraction of genetic variability for the production itself (Fontanesi *et al*., 2008).

The myostatin gene codes for a growth factor involved in muscle development. Polymorphism in this gene has economic consequences. The gene encoding myostatin was discovered in 1997 by the geneticists Alexandra Mc Pherron and Se-Jin Lee who also produced a strain of mutant mice that lack this gene. These myostatin knock-out mice apparently had twice as much muscle as normal mice, primarily due to an increased number of muscle fibres, followed by muscle cell hypertrophy and suppression of body fat accumulation (Mc Pherron *et al*., 1997; Mc Pherron and Lee, 2002). Fontanesi *et al*. (2008) detected a single nucleotide polymorphism (C-T transition) at position 34 of intron 2 of myostatin gene. In cattle, mutation identified in the coding region of the myostatin gene of Belgian Blue and Piedmontese were found to be highly associated with double muscling trait (Kambadur *et al.*, 1997; Mc Pherron *et al.*, 1997). In chicken also, mutation of myostatin gene has been reported to be associated with production traits (Gu *et al.*, 2002). In dogs, enhanced muscle development and racing performance was observed with mutated myostatin gene (Mosher *et al.*, 2007). These observations point to the fact that myostatin gene is an important candidate gene for muscle mass development in animals.

The present study focused on finding out the polymorphism of myostatin gene in New Zealand White, Soviet Chinchilla and crossbred rabbits adapted to the tropical climate of Kerala.

Materials and methods

Using phenol chloroform extraction procedure (Sambrook and Russel, 2001), DNA was isolated from blood samples collected from 60 rabbits belonging to New Zealand White, Soviet Chinchilla and crossbred rabbits maintained in the University rabbit farm of Kerala Agricultural University. The primer designed by Fontanesi *et al.* (2008) was used for the present study.

The 80 bp fragment of the myostatin gene was amplified in a thermal cycler with a final concentration of 50 ng of template DNA, 5 pmol of each primer, 2.5mM each of dNTP, 0.6 U of Taq DNA polymerase, 1X PCR buffer and 1 mM of magnesium chloride in 10 µl reaction with initial denaturation at 94 °C for 5 min, 35 amplification cycles of denaturation at 94 °C for 1 min, primer annealing at 58-61 °C for 1 min, primer extension at 72 °C for 1 min and final extension at 72 °C for 5 min. Four microlitres of the amplified product was digested overnight at 37 °C with 5U of restriction endonuclease, *AluI* and separated by 10% polyacrylamide gel electrophoresis. DNA products were visualised with ethidium bromide and the restriction pattern was analysed under UV transilluminator and documented in a gel documentation system (BioRad, Gel Doc 2000™).

Heterozygosity was calculated by the method of Ott (1992).

$$He = 1 - \sum_{i=1}^{k} p_i^2 \tag{1}$$

where p_i is the frequency of ith allele.

The Polymorphic Information Content (PIC) value was calculated using the formula:

$$PIC = 1 - \left[\sum_{i=1}^{k} p_i^2 \right] - \sum_{i=1}^{k} \sum_{j=i+1}^{k} 2 \, p_i^2 p_j^2 \tag{2}$$

where p_i and p_j are the frequencies of ith and j^{th} alleles, respectively (Botstein *et al.*, 1980).

Results and discussion

Polymerase Chain Reaction-Restriction Fragment Length Polymorphism (PCR-RFLP) revealed three restriction digestion patterns in rabbits, viz CC, CT and TT. The genotype CC was represented by a 80 bp fragment, genotype CT by three fragments of length 80 bp, 56 bp and 24 bp while TT was found represented by fragments of length 56 bp and 24 bp. Similar findings were reported in other studies.

The frequency of C allele in the pooled population was recorded as 0.57. The frequencies of C allele in New Zealand White, Soviet Chinchilla and crossbred rabbits were found to be 0.52,

0.60 and 0.60, respectively. These findings are in agreement with those of Fontanesi *et al.* (2008) who reported similar frequencies of C allele in the rabbits studied, namely Checkered Giant (0.56), Burgundy (0.60) and Giant grey (0.60). Contrary to the findings of the present study, prior research reported a higher frequency for T allele (0.6693). As regards the heterozygosity value, 0.50, 0.48, 0.48 and 0.49 were observed for New Zealand White, Soviet Chinchilla, Crossbreeds and pooled population of rabbits. The corresponding polymorphic information content (PIC) values obtained were 0.3746, 0.3648, 0.3648 and 0.3701, respectively. Similar findings have been given by Fontanesi *et al.* (2008), however lower values have been reported by others. In the present study, CT genotypes were associated with higher body weight but the difference in body weight among different genetic groups was not statistically significant. This may be due to small sample size. The study reiterates the relevance of more research activities on this front with large sample size and more DNA markers to aid association studies.

Conclusion

The present PCR-RFLP study of 80 bp fragment of myostatin gene in rabbits detected one single nucleotide polymorphism. The allele distribution of the identified polymorphism, even if it does not disrupt or affect the coding sequence because of its localisation in an intronic region without any putative functional role, seems a useful gene marker for association studies (Fontanesi *et al.*, 2008). Re-sequencing activities that are underway for a large number of animals and for the additional regions of the same gene, might provide more DNA markers in the immediate future.

Acknowledgements

The authors wish to thank the Animal Husbandry Department of Kerala state, India, for funding this project.

References

Botstein D., R.L. White, M. Skolnick and R.W. Davis, 1980. Construction of a genetic linkage map in man using restriction fragment length polymorphism. American Journal of Human Genetics, 32: 314-331.
Fontanesi L., M. Tazzoli, E. Scotti and V. Russo, 2008. Analysis of candidate genes for meat production traits in domestic rabbit breeds. 9th world Rabbit Congress, Verona, Italy.
Gu ZL., H.F. Zhang, D.H. Zhu and H. Li, 2002. Single nucleotide polymorphism analysis of the chicken Myostatin gene in different chicken lines. Yi Chan Xue Bao, 29: 599-606.
Kambadur R., M. Sharma, T.P.L. Smith and J.J. Bass, 1997. Mutations in myostatin (GDF 8) in double muscled Belgian Blue and Piedmontese cattle. Genome Research, 7: 910-915.
Mc Pherron A.C., A.M. Lawler and S.J. Lee, 1997. Regulation of skeletal muscle mass in mice by a new TGF-β super family member. Nature, 387: 83-90.
Mc Pherron A.C. and S.J. Lee, 2002. Supression of body fat accumulation in myostatin deficient mice. Journal of clinical Investigation, 109: 595-601.
Mosher D.S., P. Quignon, C.D. Bustamante, N.B. Sutter, C.S. Mellersh, H.G. Parker and E.A. Ostrander, 2007. A mutation in the myostatin gene increases muscle mass and enhances racing performance in heterozygote dogs. PLoS Genetics, 3: e79.
Ott, 1992. Strategies for characterizing highly polymorphic markers in human gene mapping. American Journal of Human Genetics, 47: 283-290.
Rothschild M.F. and M. Soller, 1997. Candidate gene analysis to detect traits of economic importance in livestock. Probe, 8: 13-22.
Sambrook J. and D.W. Russel, 2001. Molecular Cloning-A Laboratory Manuel. 3rd ed. Cold Spring Harbour Laboratory Press, New York, NY, USA, 1886 p.

Growth hormone gene in llama (*Lama glama*): characterisation and SNPs identification

M.S. Daverio, F. Di Rocco and L. Vidal Rioja
Laboratorio de Genética Molecular, Instituto Multidisciplinario de Biología Celular (IMBICE), CIC-PBA. CCT-CONICET, Calle 526 e/ 10 y 11, CP(1900), La Plata, Buenos Aires, Argentina; m.sil.daverio@hotmail.com

Abstract

The South American camelids are a source of meat, fibre, leather, transport and other products useful for the economy of populations in the north-west of Argentina. Breeding and exploitation of the llama and alpaca has experienced a notable increase and with it an interest in improving the performance of these taxa. The growth hormone (GH), a peptide hormone synthesised and secreted by the pituitary gland, affects a variety of economically important traits in livestock such as growth performance, milk production and carcass composition. In this study, we characterised the GH gene of llamas from Argentina and identified several single nucleotide polymorphisms (SNPs). As in other mammals, llama GH gene comprises five exons and four introns. Comparison with the coding sequences of the camel, whale, dolphin, pig, cow and sheep showed 99,1%, 96,5%, 96.2%, 94,8%, 90.6% and 90,3% similarity, respectively. Introns were less conserved, particularly intron C which in llama was considerably shorter than in the other compared species. Sequencing of GH gene in a sample of 20 llamas allowed the identification of 8 SNP, mainly located in introns. Nevertheless, one of them was found in the 3´region. We conclude that the polymorphisms such as the ones presented here could be used for developing genetic markers to assay the effect of different genotypes on economically important traits.

Keywords: *Lama glama*, growth hormone, nucleotide sequence, polymorphisms

Introduction

The South American camelids are represented by the wild species *Lama guanicoe* (*guanaco*) and *Vicugna vicugna* (*vicuña*) and the domestic forms *Lama pacos* (*alpaca*) and *Lama glama* (*llama*).

The llama is the largest domestic camelid widely represented in north-western Andean elevations of Argentina but also very well adapted to the central and eastern plains of the country. In the last few decades, breeding and exploitation of camelids have experienced rapid expansion that has gone beyond the borders of South American countries with a consequent increase in demand for yield and quality of their products, with fibre and meat being the most valuable ones. Herds of llama are raised in these areas with economic returns for the herders who actually look for ways to improve the efficiency of production by increasing the reproductive and growth success, the fibre quality and overcoming several adverse sanitary conditions.

The study of candidate genes, based on physiological effects, is an important tool for identifying genes to be used in marker-assisted selection programmes. GH is a peptide hormone produced and secreted by the somatotrophs of the anterior pituitary and is known for its growth-promoting activities (Vance, 1989). In most mammals it is encoded by a single gene (Forsyth and Wallis, 2002). Therefore, GH is a good candidate worth studying in livestock, since it affects a wide variety of economically important traits such as growth performance, carcass composition, and milk production.

Recently, a study in Brangus bulls population revealed that polymorphisms in GH and its transcriptional regulators appear to be predictors of growth and carcass traits in these species (Thomas *et al.*, 2007). Association of carcass traits with GH gene polymorphisms were also reported in pigs by Geldermann *et al.* (1996). In contrast, knowledge of the sequence, structure and variation of genes related to growth and reproduction in South American camelids is still an unexplored field. Hence, the aim of the present work was to determine the sequence and structure of the growth hormone gene and to identify single nucleotide polymorphisms.

Materials and methods

Samples

Blood samples were obtained from 20 randomly selected llamas belonging to two different flocks of Jujuy and La Pampa provinces, Argentina. Total genomic DNA was extracted from whole blood using standard phenol/ chloroform procedure (Sambrook and Russel, 2001).

Amplification of Llama GH gene

GH gene sequence was obtained by polymerase chain reaction (PCR). Amplification was performed in four overlapping fragments of 290-894 bp with primers (Table 1) designed on the alpaca GH mRNA (GenBank DQ782970). Primers for the amplification of 5′and 3′ends were designed in conserved sequences from GH gene of other cetartiodactyl species. PCR reactions were carried out in a programmable thermocycler (PTC-100, MJ Research) with an initial denaturation step at 94 °C for 3 min followed by 30-35 cycles of 1 min denaturation at 94 °C, 1 min annealing at 55-56 °C, 1 min chain elongation at 72 °C, and 5 min final extension at 72 °C. Amplicons were visualised in 2% agarose gels electrophoresis and stained with GelRed™ (Biotium, Hayward, California, USA). PCR products were purified and sequenced using an *automatic Genetic Analyser* 3730xl (Applied Biosystems, Foster City, USA).

Sequence analysis

Fragments obtained were assembled using software Geneious v4.7 (Drummond *et al.*, 2009) in order to determine the complete nucleotide sequence of llama GH gene. Introns, exons and translation initiation site were identified by comparison with the alpaca GH cDNA sequence from the Genebank (DQ782970). Sequence obtained, including the 5′ and 3′ regions, were aligned with related Cetartiodactyl species: dromedary (*Camelus dromedarius*, AJ575419), fin whale (*Balaenoptera physalus*, AJ831741), dolphin (*Delphinus delphis*, AJ492191), pig (*Sus scrofa*, AY727037), cow (*Bos taurus*, M57764) and sheep (*Ovis aries*, X12546) using Clustal W programme (Higgins *et al.*, 1994) followed by manual adjustment. The search for tandem repeat motifs was performed by means of Tandem Repeat Finder software (Benson, 1999).

Table 1. Primer sequences used for amplifying Llama GH gene.

Fragment	Primer Forward	Primer Reverse	T° annealing
GH 5 UTR	GAAAATAAGTGGGGGCAGAG	AGTTTCCTCCCATTATGCAG	55 °C
GH E2-4	CTGGCTGCTGACACCTACAA	ACCAGGCTGTTGGTGAAGAC	55 °C
GH I-E5	ATCCTGGGTAGCCTTCTCTC	GCACTGGAGTGGCACTTT	56 °C
GH 3 UTR	TCCTCAGGCAAACCTACGAC	TGATGCAACCTCATTTTATTAGGA	55 °C

Results and discussion

Sequence and organisation of llama GH gene

The obtained sequence consisted of 1781 bp that contain a short fragment of the 5′UTR (5′untranslated region), the complete llama GH gene and 88 bp of the 3′UTR region. The whole sequence was deposited at GenBank database under the accession number HM921333. The position +1 of the sequence was assigned coinciding with the first nucleotide of the start codon ATG.

As in other mammals (Barta *et al*., 1981; Wallis M, 2008) the exon-intron organisation consists of five exons (1-5) disrupted by four introns (A to D) (Figure 1). The potential polyadenylation site AATAAA is located 80 bp 3′ from stop codon (TAG). All of the intron-exon junctions conform to the consensus splice sequence consisting of the dinucleotides GT and AG at the 5′ and 3′ boundaries respectively, (Breathnach and Chambon, 1981) allowing the prediction of a 216 amino acid polypeptide.

Exons of the GH gene are highly conserved between cetartiodactyls, both in length and sequence. Exon 1 whose region from -60 to -1 is transcribed but not translated, was the most variable. Comparison of the coding region with those of dromedary, whale, dolphin, pig, cow and sheep showed 99,1%, 96,5%, 96.2%, 94,8%, 90.6% and 90,3% similarity.

As expected, introns are less conserved, especially intron C which in llama is shorter than those of all species compared. It contains two imperfect repeats of 21bp, each spanning the following sequences: 5′GGCGGCGGCGGGTGATGGGAT3′ and 5′GGCGGCGGAGGATGGTGGGTT3′.

The deduced amino acid sequence of llama GH is identical to the alpaca protein and differs from the dromedary by a V9M replacement within the signal peptide (Figure 2). In concordance with the observations for GH nucleotide sequence, the llama protein showed a higher homology with dolphin, whale and pig (96-97%) than with sheep and cow (87-88%). These results can be explained by the burst of rapid evolution of GH that occurred after the separation of the line leading to ruminants from other cetartiodactyls (Maniou *et al*., 2004)

Interesting to note is a T/C substitution in the 5′ region at the first base of the TATA box that llama share with the dromedary but which differs from the conserved TATAAA sequence seen in other cetartiodactyl species (Maniou *et al*., 2004). However, this change does not seems to affect GH expression since Martinat *et al*. (1990) have isolated the dromedary protein from pituitary.

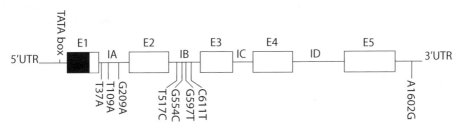

Figure 1. Diagram of the structural organisation of llama GH gene and SNPs distribution. Exons (E1-E5) are shown as boxes and introns and flanking regions are shown as bars. Black area: untranslated gene sequences.

Signal peptide
+1

MAAGPRTSMLLAFTLLCLPWPQEAGAFPAMPLSSLFANAVLRAQHLHQLAADTYKEFERTYIPE
GQRYSIQNAQAAFCFSETIPAPTGKDEAQQRSDVELLRFSLLLIQSWLGPVQFLSRVFTNSLVFGT
SDRVYEKLKDLEEGIQALMRELEDGSPRAGQILRQTYDKFDTNLRSDDALLKNYGLLSCFKKDL
HKAETYLRVMKCRRFVESSCAF

Figure 2. Sequence of the llama GH protein. The bar represents the signal peptide, amino acid +1 represents start site of the mature protein.

Evidence of llama GH gene variability is sustained by the discovery of 8 SNPs in the complete sequence of the 20 analysed samples. Most of the identified SNPs were located in non-coding regions and mainly concentrated in Introns A and B (Figure 1). Intron C and D, as well as exons, were invariable. This study also included a 88 bp segment of the 3′ UTR region of the GH gene where one additional SNP was detected.

Although non coding SNPs are considered less important than the coding ones, recent studies in other species demonstrated the association of intronic SNPs with growth and carcass traits (Nie *et al.*, 2005; Thomas *et al.*, 2007).

Conclusion

This work presents the first complete sequencing and organisation of the llama GH gene in addition to the identification of several allelic variants. The polymorphisms such as the ones identified here will be useful for future genotype-phenotype association studies.

Acknowledgements

We gratefully acknowledge Miriam B. Silbestro for her technical assistance.

References

Barta A., R.I. Richards, J.D. Baxter and J. Shine, 1981. Primary structure and evolution of rat growth hormone gene. Proceedings of the National Academy of Sciences, 78: 4867-4871.

Benson G., 1999. Tandem repeats finder: a program to analyze DNA sequences. Nucleic Acids Research, 27: 573-580.

Breathnach R. and P. Chambon, 1981. Organization and expression of eukaryotic split genes coding for proteins. Annual Review Biochemistry, 50: 349-383.

Drummond A.J., B. Ashton, M. Cheung, J. Heled, M. Kearse, R. Moir, S. Stones-Havas, T. Thierer and A. Wilson, 2009. Geneious v4.7. Available at http://www.geneious.com.

Forsyth I.A. and M. Wallis, 2002. Growth hormone and prolactin-molecular and functional evolution. Journal of Mammary Gland Biology and Neoplasia, 7: 291-312.

Geldermann H., E. Muller, P. Beeckmann, C. Knorr, G. Yue and G. Moser, 1996. Mapping of quantitative Polymorphism of *PIT1*, *GH*, and *GHRH* genes and economic traits of pigs 199 trait loci by means of marker genes in F2 generations of wild boar, Pietrain and Meishan pigs. Journal of Animal Breeding and Genetics, 113: 381-387.

Higgins D., J. Thompson, T. Gibson, J.D. Thompson, D.G. Higgins and T.J. Gibson, 1994. CLUSTAL W: improving the sensitivity of progressive multiple sequence alignment through sequence weighting, position-specific gap penalties and weight matrix choice. Nucleic Acids Research, 22: 4673-4680.

Maniou Z., O.C. Wallis and M. Wallis, 2004. Episodic molecular evolution of pituitary growth hormone in Cetartiodactyla. Journal of Molecular Evolution, 58: 743-753.

Martinat N., A. Anouassi, J.C. Huet, J.C. Pernollet, V. Segard and Y. Combarnous, 1990. Purification and partial characterization of growth hormone from the dromedary (*Camelus dromedarius*). Domestic Animal Endocrinology, 7: 527-536.

Nie, Q., B. Sun, D. Zhang, C. Luo, N.A. Ishag, M. Lei, G. Yang and X. Zhang, 2005. High diversity of the chicken growth hormone gene and effects on growth and carcass traits. Journal of Heredity 96: 698-703.

Sambrook J. and D.W. Russel, 2001. Preparation and analysis of eukaryotic genomic DNA. Molecular Cloning: A laboratory manual. 3rd ed, Cold Spring Harbor Laboratory Press, New York, NY, USA.

Thomas M.G., R.M. Enns, K.L.Shirley, M.D. Garcia and A.J. Garret, 2007. Associations of DNA polymorphisms in growth hormone and its transcriptional regulators with growth and carcass traits in two populations of Brangus bulls. Genetics and Molecular Research, 6: 222-237.

Vance M.L., 1989. Growth hormone – metabolic effect. Frisch, H. and Thorner M.O. Hormonal Regulation of Growth. Serono Symposia Publications. Raven Press. New York, 58: 201-226.

Wallis M., 2008. Mammalian genome projects reveal new growth hormone (GH) sequences. Characterization of the GH-encoding genes of armadillo (*Dasypus novemcinctus*), hedgehog (*Erinaceus europaeus*), bat (*Myotis lucifugus*), hyrax (*Procavia capensis*), shrew (*Sorex araneus*), ground squirrel (*Spermophilus tridecemlineatus*), elephant (*Loxodonta africana*), cat (*Felis catus*) and opossum (*Monodelphis domestica*). General and Comparative Endocrinology, 155: 271-279.

Alpaca sperm chromatin evaluation using Toluidine Blue

M.I. Carretero[1], C.C. Arraztoa[1], C.I. Casaretto[1], W. Huanca[2], D.M. Neild[1] and M.S. Giuliano[1]
[1]Instituto de Investigación y Tecnología en Reproducción Animal (INITRA), Facultad de Ciencias Veterinarias, Universidad de Buenos Aires, Argentina; ignaciacarretero@fvet.uba.ar
[2]Laboratorio de Reproducción Animal, Facultad de Medicina Veterinaria, Universidad Mayor de San Marcos, Perú

Abstract

Alpacas are a domestic species of South American camelids (SAC), which are internationally appreciated because they have a high-quality fibre. Toluidine Blue (TB) is a cationic stain that binds to DNA permitting differentiation between sperm heads according to the degree of chromatin decondensation. This technique was used to evaluate sperm chromatin in alpacas, determine chromatin condensation patterns in alpacas and determine if it is possible to use dithiothreitol (DTT) as a positive control for the stain. A total of 15 ejaculates were collected from 9 alpacas using electroejaculation. TB stain was carried out according to Carretero *et al.* (2009) on smears of raw semen. Spermatozoa were classified according to the degree of chromatin decondensation. Descriptive statistics and analysis of variance for evaluating male differences were performed. Three patterns of staining with TB were observed in alpaca sperm: light blue (negative, without alteration of chromatin condensation), light violet (intermediate, some degree of decondensation), dark blue-violet (positive, high degree of decondensation). The percentages observed were (mean ± SD): TB positive 7.55±5.22, TB intermediate 21.67±6.81 and TB negative 70.78±10.80. In the spermatozoa incubated with 1% DTT, 2 main categories were observed: reacted (positive with TB) and non-reacted (negative with TB). In reacted sperm, 3 sub-categories were observed according to sperm head morphology and presence of vacuoles. Variability between males was observed for positive ($P=0.03$) but not for intermediate TB stained sperm. It is possible to use TB to evaluate the degree of chromatin decondensation in alpaca spermatozoa, showing the same TB patterns as in other SAC; also 1% DTT can be used as a positive control for this technique.

Keywords: sperm DNA, Toluidine Blue, alpacas

Introduction

Alpacas are a domestic species of South American camelids (SAC), which are internationally appreciated because they have a high-quality fibre and meat with low cholesterol content. Therefore all reproductive techniques that contribute to genetic selection are very important in these species. Standard semen evaluation includes characteristics such as ejaculate volume, sperm concentration, plasma membrane function and integrity, sperm motility and sperm morphology. However, there is now evidence to support the idea that the integrity of sperm chromatin is important not only for fertilisation but also for normal embryonic development (Ward *et al.*, 2000; Morris *et al.*, 2002). Toluidine Blue (TB) is a nuclear dye used to evaluate the degree of sperm chromatin condensation by detecting the absence or rupture of disulfide bonds; thus, stained nuclei (metachromatic blue-violet colouring) indicate altered chromatin condensation. Using well-defined groups of fertile and infertile male, this stain was used to evaluate male fertility potential (Tsarev *et al.*, 2009). In addition, TB is simple and inexpensive; all that is needed is a light microscope and the dyes. In our lab, this stain has been used to evaluate the degree of DNA decondensation and to determine the stain patterns in raw semen of llamas (Carretero *et al.*, 2009) and guanacos (Carretero *et al.*, 2010). Also, in llama and guanaco, 1% dithiothreitol (DTT) was effective when used as a positive control of the TB stain because this is an agent which reduces disulfide bonds. The objectives of this study were to see if this technique can be used for evaluating alpaca sperm chromatin, determining the chromatin condensation patterns in this species using TB and determining if it is possible to use 1% DTT as a positive control for the stain.

Materials and methods

Animals and location

The study was carried out at the Faculty of Veterinary Medicine of the University of San Marcos, Lima, Peru and at the Faculty of Veterinary Sciences of the University of Buenos Aires, Argentina. A total of 9 male *Lama pacos* were used. All animals had a good nutritional status (body condition) and were healthy at the time of the trial (February).

Semen collection

A total of 15 ejaculates were collected from 9 alpacas using electroejaculation according to the technique described by Director *et al.* (2007).

DNA evaluation

TB stain was carried out according to Carretero *et al.* (2009) on smears of raw semen. Briefly, samples were smeared on clean, non-greasy slides and once dry, fixed with ethanol 96° for 2 minutes. Smears were stained for 5 minutes with a working solution of TB and then rinsed with distilled water and air dried, protected from the light. The TB working solution was prepared by diluting a stock solution (0.4 grams in 200 ml distilled H_2O) 1:9 (1 part stock solution in 9 parts buffer, pH 4). Preparations were observed directly under immersion oil using a phase contrast microscope (1000X) evaluating a minimum of 200 spermatozoa per smear. Spermatozoa were classified according to the degree of chromatin decondensation. 1% Dithiothreitol (DTT), a substance that reduces chromatin disulphide bonds, was used as a positive control for the stain. Equal quantities of sample and 1% DTT were placed in Eppendorfs and the reaction was allowed to progress at room temperature for 3 minutes; after this the smears were made and blow dried to prevent the reaction continuing on the slides. Then the smears were fixed and stained with TB according to the previously described protocol. Sperm were evaluated with a phase contrast microscope (minimum 200 spermatozoa) at 1000X and were classified in 2 main categories according to chromatin colouring and in sub-categories according to sperm head morphology and presence of vacuoles.

Statistical analysis

Descriptive statistics and analysis of variance for evaluating male differences were performed.

Results

According to the degree of chromatin decondensation, three patterns of staining with TB were observed in alpaca sperm: light blue (negative, without alteration of chromatin condensation), light violet (intermediate, some degree of decondensation), dark blue-violet (positive, high degree of decondensation). The percentages observed were (mean ± SD): TB positive 7.55±5.22, TB intermediate 21.67±6.81 and TB negative 70.78±10.80. In the spermatozoa incubated with 1% DTT, 2 main categories were observed: reacted (positive with TB) and non-reacted (negative with TB). In reacted sperm, 3 sub-categories were observed: (1) very decondensed heads (extremely large heads with lots of vacuoles inside), (2) deformed heads (large heads with a moderate number of vacuoles inside) and (3) heads with normal shape and size. In addition, variability between males was observed for positive ($P=0.03$) but not for intermediate TB stained sperm.

Discussion

The TB stain has been used in various species to evaluate sperm chromatin condensation (Erenpreisa *et al.*, 2003; Beletti and Mello, 2004; Beletti *et al.*, 2005; Sardoy *et al.*, 2008) and also in llama and guanaco (Carretero *et al.*, 2009, 2010), but has not been reported in alpaca sperm.

In this study we observed the same 3 patterns of TB staining as those observed in llama, guanaco and equine sperm (Sardoy *et al.*, 2008; Carretero *et al.*, 2009; Carretero *et al.*, 2010). Erenpreisa *et al.* (2003) reported 4 patterns of TB staining in human sperm and although we were able to distinguish the fourth pattern these authors describe, the evaluation becomes much more subjective. With regard to the variability observed between alpaca males in the TB positive sperm, this observation coincides with that found in other species such as llama, human and bull (Bedford *et al.*, 1973; Pasteur *et al.*, 1983; Vieytes *et al.*, 2008; Carretero *et al.*, 2009).

The mean percentage of TB positive + intermediate sperm in alpaca (29.21±10.79) was similar to TB positive + intermediate llama sperm observed in ejaculates collected in summer (25.82±13) and was higher than that observed in llama sperm collected during the winter season (TB positive + intermediate: 13.18±8.02). Llama sperm showed significant differences in DNA condensation between the winter and summer seasons (Carretero *et al.*, 2008) and these preliminary data would seem to indicate that the alpaca perhaps evidence the same differences. This would need to be corroborated with further studies. In guanaco ejaculates, collected during April (beginning of autumn), the mean percentage of TB positive + intermediate sperm (37.2±9.6) was higher than that observed in alpaca and llama sperm. This difference might be because the guanaco showed sub-optimal body conditions at the time of the study due to low nutritional conditions.

The patterns we observed in alpaca spermatozoa incubated with 1% DTT were exactly the same as those observed in llama, guanaco and equine sperm when incubating samples with the reducing agent for 3 minutes. Similar patterns, but in different proportions, have been observed in other species (rabbit: Beil and Graves, 1977; ram: Rodriguez *et al.*, 1985; human: Barrera *et al.*, 1993; and bull: Beil and Graves, 1977; Vieytes *et al.*, 2008). In those studies authors used higher incubation times and different DTT concentrations. Therefore, the use of a reducing agent as a positive control should be adapted to each species.

Conclusions

It is possible to use TB to evaluate the degree of chromatin decondensation in alpaca spermatozoa, showing the same TB patterns as in other SAC; also 1% DTT can be used as a positive control for this technique. Toluidine Blue staining will be useful for evaluating the effects that different reproductive biotechnologies have on alpaca sperm DNA condensation.

Acknowledgements

The project was supported by the Grant N° 064-FINCyT-PIBAP-UNMSM 2008. The authors wish to thank María Chileno, Elvira Cano, Diana Tolentino, Lenin Benavides for their assistance in the project.

References

Barrera C., A.B. Mazolli, C. Pelling and C. Stockert, 1993. Metachromatic staining of human sperm nuclei after reduction of disulphide bonds. Acta Histochemica, (Jena), 94: 141-149.

Bedford J.M., H.I. Calvin and G.W. Cooper, 1973. The maturation of spermatozoa in the human epididymis. Journal of Reproduction and Fertility, 18 (Suppl.): 199-213.

Beil R.E. and C.N. Graves, 1977. Nuclear decondensation of mammalian spermatozoa: changes during maturation and *in vitro* storage. Journal of Experimental Zoology, 202: 235-40.

Beletti M.E. and M.L.S. Mello, 2004. Comparison between the toluidine blue stain and the feulgen reaction for evaluation of rabbit sperm chromatin condensation and their relationship with sperm morphology. Theriogenology, 62: 398-402.

Beletti M.E., L. da Fontoura Costa and M. Mendes Guardieiro, 2005. Morphometric features and chromatin condensation abnormalities evaluated by toluidine blue staining in bull spermatozoa. Brazilian Journal of Veterinary Research and Animal Science, 22: 85-90.

Carretero M.I., S. Giuliano, C. Casaretto, M.Gambarotta and D. Neild, 2008. Influencia de la estación del año sobre el ADN de espermatozoides de llama. Invet, 10: 150.

Carretero M.I., S.M. Giuliano, C.I. Casaretto, M.C. Gambarotta and D.M. Neild, 2009. Evaluación del ADN espermático de llamas utilizando azul de toluidina. Invet, 11: 55-63.

Carretero M.I., S. Giuliano, A. Agüero, M. Pinto, M. Miragaya, V. Trasorras, J. Egey, J. von Thungen and D. Neild, 2010. Guanaco sperm chromatin evaluation using Toluidine Blue. Reproduction, Fertility and Development, 22: 310-310.

Director A., S. Giuliano, V. Trasorras, M.I. Carretero, M. Pinto and M. Miragaya, 2007. Electroejaculation in llama (*Lama glama*). Journal of Camel Practice and Research, 14: 203-206.

Erenpreisa J., J. Erenpreiss, T. Freivalds, M. Slaidina, R. Krampe, J. Butikova, A. Ivanov and D. Pjanova, 2003. Toluidine blue test for sperm DNA integrity and elaboration of image cytometry algorithm. Cytometry Part A, 52A: 19-27.

Morris I.D., S. Holt, L. Dixon and D.R. Brison, 2002. The spectrum of DNA damage in human sperm assessed by single-cell gel electrophoresis (Comet assay) and its relationship to fertilisation and embryo development. Human Reproduction, 17: 990-998.

Pasteur X., J.L. Laurent, J. Azéma, M. Jourlin, P. Deage and D. Grange, 1983. Decondensation *in vitro* du noyau du spermatozoide humain et analyse d'image. Comptes Rendus des Seances de la Societe de Biologie et de ses filiales, 176: 123.

Rodriguez H., C. Ohanian and E. Bustos-Obregon, 1985. Nuclear chromatin decondensation of spermatozoa *in vitro*: a method for evaluating the fertilizing ability of ovine semen. International Journal of Andrology, 8: 147-158.

Sardoy M.C., M.I. Carretero and D. Neild, 2008. Evaluation of stallion sperm DNA alterations during cryopreservation using toluidine blue. Animal Reproduction Science, 107: 349-350.

Tsarev I., M. Bungum, A. Giwercman, J. Erenpreisa, T. Ebessen, E. Ernst and J. Erenpreiss, 2009. Evaluation of male fertility portential by toluidine blue test for sperm chromatin structure assessment. Human Reproduction, 24: 1569-1574.

Vieytes A.L., H.O. Cisale and M.R. Ferrari, 2008. Relationship between the nuclear morphology of the sperm of 10 bulls and their fertility. Veterinary Record, 163: 625-629.

Ward W.S., H. Kishikawa, H. Akutsu and R. Yanagimachi, 2000. Further evidence that sperm nuclear proteins are necessary for embryogenesis. Zygote, 8: 51-56.

Derivation of economic values for fibre diameter fleece weight in alpacas

L. Alfonso[1], J. Buritica[1], R. Quispe[2] and I. Quicaño[2]
[1]Universidad Pública de Navarra, Pamplona, Spain; leo.alfonso@unavarra.es
[2]Desco Huancavelica, Huancavelica, Peru

Abstract

Economic values for fibre diameter and fleece weight in alpacas were derived using profit equations. Parameters in the profit equation were chosen to reflect the production and commercialisation of alpaca fibre in Huancavelica (Peru). Income depends on the weight and colour of fleeces commercialised from each of the four different categories defined by the technical Peruvian normative: Extra fine (EF), Fine (F), Semi Fine (SF) and Coarse (C). Basically, the four categories are determined by the amount of fibre in the whole fleece finer than 26.5 µm (>70%, 69-55%, 55-40%, <40%, respectively). The selection criteria to improve the quality and quantity of fibre produced are respectively the fibre diameter in the mid-side site (DF) of alpacas and the greasy fleece weight (PV). DF was assumed as an underlying continuous variable and threshold values for the four quality categories (EF, F, SF and C) were approximated using the relationship between DF and the fibre diameter in the whole fleece. Economic weights were derived by first derivatives of profit equation with respect DF and PV considering costs uncorrelated with both parameters. The prices paid in the region of Huancavelica for the different categories and colours (white, cream, other colours) in the period 2005-2009 were considered to obtain economic values and to analyse the effect of changes in fibre price. Economic values were greater for PV than for DF, and greater for white and cream alpacas than for alpacas with other colours. Sensitivity analysis of economic values to changes in price showed that they vary considerably, and also the ratio of economic values varies for DF and PV. The DF seems to be economically more important when prices paid to producers are lower. Nevertheless, results should not be applied in breeding programmes without other analyses being performed using other participatory approaches.

Keywords: *Vicugna pacos*, quality, selection, Andean

Introduction

In Huancavelica, situated in the Peruvian *Andean Plateau*, the small herders of alpacas commercialise their fibre through middlemen from big textile enterprises or participating associations. In the latter situation, the entire fleeces are carried to a store centre after shearing and categorised following the 231.300 Peruvian Technical Standards (NTP, 2004). These standards indicate four different categories determined by its percentage of highest quality fibres (<26.5 µm): Extra Fine (70% minimum of highest quality fibres), Fine (between 55 and 69%), Semi Fine (between 40 and 55%) and Coarse (less than 40%).

After categorisation, the fleeces are bought from producers according to their category and they move to the following step of the fibre chain commercialisation and transformation: the classification according to the 231.301 Peruvian Technical Standards (NTP, 2004). The fibre of fleeces is separated according to its diameter in five classes: Baby (<23 µm), Fleece (between 23.1 and 26.5), Medium Fleece (between 26.6 and 29 µm), Huarizo (between 29.1 and 31.5) and Coarse (>31.5). The categorisation and classification are made in situ by trained women.

This commercialisation model implies that producers' incomes (I) when selling the fibre are not determined by the amount of fibre of each of the five classes produced (Baby, Fleece, Medium Fleece, Huarizo, Coarse), but by the amount of fleeces sold from each of the four categories (Extra Fine, Fine, Semi Fine, Coarse). So, prices paid for fleeces depend on weight, category, alpaca type (Huacaya or Suri), and colour (white, cream or other colours). The costs (C) can be considered independently from the quantity and quality of fibre produced, because of the production system in which alpaca production is based in this region (Montes *et al.*, 2008).

Different initiatives have been carried out in the Huancavelica region to genetically improve alpaca fibre production. Taking into account the payment system, the main breeding objectives are the quantity and the quality of fibre. As in other animal species producing high-quality fibre, the total weight and the diameter are the two traits with most economic relevance. To evaluate both objectives, the greasy fleece weight and the mean fibre diameter (measured in a sample from the midsize site of alpacas) are considered as selection criteria (Quispe *et al.*, 2009).

The selection for two or more traits can be done in breeding programmes using a linear function by weighting the genetic value of each trait by its economic value. The resulting value, one for each animal, is the value to be considered for selecting future reproducers. Till now, in the Huancavelica region, economic weights to be applied to greasy fleece weight and mean fibre diameter have not been analysed in detail.

The aim of this work was, first, to analyse prices paid for the fibre of Huacaya alpaca to producers in the Huancavelica region during the last five years and according to the quality and colour, and second, to approximate the economic values for greasy fleece weight and mean fibre diameter taking into account the prices paid during that period.

Material and methods

Information about prices paid for Huacaya fibre to alpaqueros in the Huancavelica region were collected from APROAL (Asociación de Promotores Alpaqueros de Huancavelica). Data were computed for different quality categories, fibre colour and commercial campaigns. Prices were not always fixed for the different campaigns throughout the year, sometimes varying considerably (see Table 1).

Profit equations were used to calculate economic values by partial differentiation to fibre diameter and fleece weight. Because of a payment system for fibre quality that corresponds to a categorical trait, the NTP (2004) categories, an underlying continuous variable was assumed to compute partial derivatives (as in Bekman and Van Arendonk, 1993).

Assuming a normal distribution for mean fibre diameter, as found in previous works (Montes *et al.*, 2008), and considering the variability estimated in the whole fleece by Aylan-Parker and McGregor (2002) threshold values were computed for the mean fibre diameter of each category (EF, Extra Fine; F, Fine; SF, Semi Fine; C, Coarse) in the whole fleece. First, the three threshold values that determine the four categories were calculated in the standard normal distribution: 0.524, 0.126 and -0.253. Second, taking into account that they should correspond to an average value of 26.5 µm and assuming an estimated standard deviation of 5.3 µm (Aylan-Parker and McGregor, 2002), they were translated to mean fibre diameter value in the whole fleece: 23.72, 25.83 and 27.84 µm. Finally, using the regression equation estimated by Aylan-Parker and McGregor (2002) relating mean fibre diameter in the whole fleece (DFV) and in the midsize site sampling point, used as selection criterion, (DF) [DFV=2.70+1.01DF] threshold values were established as 20.81, 22.90 and 24.89 µm.

Table I. Prices (in soles/lb) paid by APROAL in different campaigns for the categorised fibre to producers according to quality and colour.

Campaign	White				Cream				Other colours			
	EF[1]	F	SF	C	EF	F	SF	C	EF	F	SF	C
2005	10.9	9.1	6.9	3.5	9.4	7.9	5.7	2.7	5.9	4.6	3.4	2.4
2006-1st	11.1	9.3	7.6	4.0	9.3	7.8	6.0	3.0	5.8	4.8	3.8	2.8
2006-2nd	11.8	9.8	8.0	4.3	9.5	8.2	6.4	3.2	6.1	5.1	4.0	3.0
2006-3st	12.3	10.3	8.2	4.8	9.8	8.5	7.1	3.4	6.1	5.1	4.0	3.0
2007-1st	13.4	11.2	9.3	5.1	10.4	8.6	7.4	3.1	6.6	5.1	4.1	2.8
2007-2nd	13.8	11.6	9.7	5.5	10.8	9.0	7.8	3.5	7.0	5.5	4.5	3.2
2007-3st	13.6	11.3	9.3	6.0	10.4	8.6	7.4	4.0	6.6	5.1	4.2	3.0
2008	9.0	8.0	6.0	4.0	8.0	7.0	5.0	3.5	4.0	3.0	2.0	2.5
2009-1st	7.5	5.5	3.5	3.0	5.5	4.0	2.5	1.8	5.0	3.5	2.5	1.5
2009-2nd	9.0	7.0	6.0	4.5	6.0	5.0	4.0	3.0	6.0	5.0	4.0	3.0
Mean	11.24	9.31	7.45	4.47	8.91	7.46	5.93	3.12	5.91	4.68	3.65	2.72
sd	2.17	1.99	1.91	0.91	1.84	1.67	1.69	0.59	0.86	0.80	0.80	0.49

[1] Categories: EF, Extra Fine; F, Fine; SF, Semi Fine; C, Coarse.

Once the categorical variable was transformed, the economic weightings for mean fibre diameter in the sampling point (DF) and for greasy fleece weight (PV) were computed. Assuming that the fibre production cost is independent of both selection criteria, the calculation was simplified to partial derivation of income function. So, for each colour and storing campaign, the income will be determined by:

$$I = \Phi(t_1-\mu) \cdot p_1 \cdot s_1 + [\Phi(t_2-\mu)-\Phi(t_1-\mu)] \cdot p_2 \cdot s_2 + [\Phi(t_3-\mu)-\Phi(t_2-\mu)] \cdot p_3 \cdot s_3 + [1-\Phi(t_3-\mu)] \cdot p_4 \cdot s_4 \quad (1)$$

where:
$\Phi(t)$ is the accumulated standard normal distribution,
$(t_i-\mu)$ is the distance between the mean fibre diameter in the sampling point and the t threshold value of i category in units of the underlying standard normal scale,
p_i is the weight, in pounds, of a fleece from i category,
s_i is the amount of 'soles' paid by pound of a fleece from i category,

Considering DF and PV as independent, $PV=p_1=p_2=p_3=p_4$, (although related, relationship is low (Quispe et al. 2009)) and given that the derivative of $\Phi(x)$ is the density function $[\Phi'(x)=\varphi(x)]$, the expression that determines the economic value for DF was:

$\partial I/\partial DF =$
$-\varphi(t_1-\mu) \cdot p_1 \cdot s_1 - \varphi(t_2-\mu) \cdot p_2 \cdot s_2 + \varphi(t_1-\mu) \cdot p_2 \cdot s_2 - \varphi(t_3-\mu) \cdot p_3 \cdot s_3 + \varphi(t_2-\mu) \cdot p_3 \cdot s_3 + \varphi(t_3-\mu) \cdot p_4 \cdot s_4 =$
$\varphi(t_1-\mu) \cdot (p_2 \cdot s_2 - p_1 \cdot s_1) + \varphi(t_2-\mu) \cdot (p_3 \cdot s_3 - p_2 \cdot s_2) + \varphi(t_3-\mu) \cdot (p_4 \cdot s_4 - p_3 \cdot s_3) =$
$pv \cdot [\varphi(t_1-\mu) \cdot (s_2-s_1) + \varphi(t_2-\mu) \cdot (s_3-s_2) + \varphi(t_3-\mu) \cdot (s_4-s_3)]$ (2)

Similarly, the expression that determines the economic value for PV was:

$\partial I/\partial PV =$
$\Phi(t_1-\mu) \cdot s_1 + [\Phi(t_2-\mu)-\Phi(t_1-\mu)] \cdot s_2 + [\Phi(t_3-\mu)-\Phi(t_2-\mu)] \cdot s_3 + [1-\Phi(t_3-\mu)] \cdot s_4 =$

$$\Phi(t_1-\mu)\cdot(s_1-s_2) + \Phi(t_2-\mu)\cdot(s_2-s_3) + \Phi(t_3-\mu)\cdot(s_3-s_4) + s_4 \qquad (3)$$

Values for t_i, $\Phi(t_i-\mu)$ y $\varphi(t_i-\mu)$ for the NTP (2004) categories in each fibre colour class (white, cream and other colours) were obtained considering mean and standard deviation values estimated by Oria *et al.* (2009). The average fleece weight was considered of 5 lb following the estimations of Quispe *et al.* (2009) in white Huacayas. For cream and other colours animals the same weight was assumed because of the lack of estimates in the region.

Results and discussion

Table 1 shows the prices of Huacaya alpaca fibre paid by APROAL in the period 2005-2009. First, the important variations between colours and quality categories should be noted. The white fibre and Extra Fine and Fine categories are the most valued. In some campaigns the price paid for the coarse white fibre is similar to that paid for the Extra Fine coloured fibre (2008, Table 1). Second, the variations between years, and even campaigns of the same year, are also very important. As an example, the maximum difference observed for the more valuable fibre (Extra Fine-White) was 6.3 soles/lb, a variation of 85% with respect to the lowest price. In this scenario it is necessary to analyse carefully the economic weightings of quality and quantity of fibre produced.

The derived economic weightings are shown in Table 2. They represent the difference in profit per animal between the previous and posterior situations of the genetic improvement of DF and PV (i.e. the marginal profit for the diminution of DF in a μm, and the augmentation of PV in a pound). So, when DF is decreased a μm the profit is increased between 5.4 and 8.0 soles per white alpaca, between 3.3 and 6.3 per cream alpaca, and between 2.0 and 2.5 per alpaca with another colour. Similarly, increasing the fleece weight by a pound implies increasing the profit between 5.1 and 10.5 soles per white alpaca, between 3.3 and 6.3 per cream alpaca, and between 2.2 and 4.0 per alpaca with other colour. These values show the differences in the profitability of selecting white alpacas rather than coloured alpacas, especially with colours different from cream.

Using the ratio between the economic values of diameter and weight, it can be observed that the increase by a pound of weight fleece is more profitable than the improvement of mean fibre diameter by its reduction in a μm. This is especially true in coloured alpacas, but not so clear in white and cream alpacas. Nevertheless, the results confirm the economic interest in decreasing the fibre diameter, as showed in other works carried out in alpacas (McGregor, 2006).

However, to interpret these results properly in terms of the selection interest of quality *vs.* quantity, other parameters should be considered. Quispe *et al.* (2009) estimated the expected genetic progress under different scenarios, and found figures ranging from -0.17 and -0.35 μm, for DF and 0.15 and 0.29 lb for PV. Following these results, the profit increase could be between 0.4 and 2.5 soles per animal/year for fibre quality selection, and between 0.5 and 2.4 soles for fibre quantity selection.

One last remark concerns the relationship between the relevance of selection for fibre quality with an economic context. When the fibre price increases in the regional market of Huancavelica, the interest in producing fibre of good quality is reduced, especially for the more valuable colours (Figure 1). Selection for finest fibre seems to be less profitable in a good economic context, but less risky because of the huge variations observed in prices paid to alpaca producers.

Table 2. Economic values for mean fibre diameter (μm) and greasy fleece weight (lb) according to colour and campaign (in soles).

	White			Cream			Other colours		
Campaign	Diameter	Weight	Ratio[1]	Diameter	Weight	Ratio	Diameter	Weight	Ratio
2005	-7.50	7.89	0.95	-6.25	5.67	1.10	-2.46	3.10	0.79
2006-1st	-6.52	8.27	0.79	-5.49	5.82	0.94	-2.00	3.45	0.58
2006-2nd	-7.06	8.76	0.81	-5.22	6.10	0.86	-2.14	3.67	0.58
2006-3st	-7.65	9.19	0.83	-4.45	6.45	0.69	-2.14	3.67	0.58
2007-1st	-7.61	10.07	0.76	-4.75	6.53	0.73	-2.32	3.60	0.65
2007-2nd	-7.61	10.47	0.73	-4.75	6.93	0.68	-2.32	4.00	0.58
2007-3st	-7.97	10.35	0.77	-4.74	6.88	0.69	-2.18	3.74	0.58
2008	-5.71	6.96	0.82	-5.19	5.38	0.96	-1.99	2.56	0.78
2009-1st	-7.40	5.08	1.46	-4.90	3.09	1.59	-2.32	2.18	1.06
2009-2nd	-5.44	6.81	0.80	-3.27	4.19	0.78	-2.00	3.65	0.55
Average [2]	-7.05	8.38	0.87	-4.90	5.70	0.90	-2.19	3.36	0.67

[1] Diameter economic value / Weight economic value.
[2] Computed with the average prices paid for the entire period.

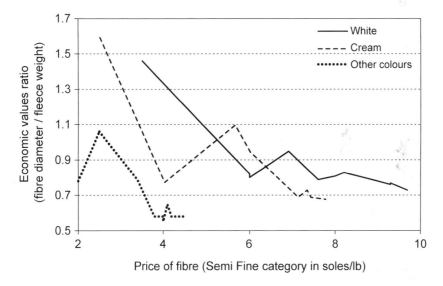

Figure 1. Relationship between the price of fibre (taking as reference the category Semi Fine) and the relevance of selection for mean fibre diameter with respect the greasy fleece weight (ratio between its economic values).

Conclusion

The results indicate a high variability in the price of alpaca fibre in the regional market of Huancavelica. Therefore, the derivation of robust economic values for the quality and quantity of fibre is risky in the medium to long term needed by a breeding programme. However, from a conservative point of view and considering a low-price scenario, the selection for fibre quality is clearly justified. The profit increase expected for selection in white Huacaya alpacas would more easily justify the breeding programme cost that in cream or other coloured Huacaya alpacas. The economic values derived in the present work should be interpreted with caution and they cannot be directly employed in a breeding programme in the region because it would be desirable to first use other methodologies like that employed by Gizaw *et al*. (2010) based on a participative definition of breeding goals.

Acknowledgements

Thanks to APROAL for providing the data on prices and amounts of fibre they commercialised in the Huancavelica region.

References

Aylan-Parker J. and M. McGregor, 2002. Optimising sampling techniques and estimating sampling variance of fleece quality attributes in alpacas. Small Ruminant Research, 44: 53-64.

Bekman H. and J.A.M. van Arendonk, 1993. Derivation of economic values for veal, beef and milk production traits using profit equations. Livestock Production Science, 34: 35-56.

Gizaw S., H.Komen and J.A.M. van Arendonk, 2010. Participatory definition of breeding objectives and selection indexes for sheep breeding in traditional systems. Livestock Science, 168: 67-74.

McGregor B.A., 2006. Production, attributes and relative value of alpaca fleeces in southern Australia and implications for industry development. Small Ruminant Research, 61: 93-111.

Montes M., I. Quicaño, R. Quispe, E. Quispe and L. Alfonso, 2008. Quality characteristics of Huacaya alpaca fibre produced in the Peruvian Andean Plateau region of Huancavelica. Spanish Journal of Agricultural Research, 6: 33-38.

NTP, 2004. Normas técnicas peruanas 231.300, 231.301 (Peruvian technical standards). INDECOPI.

Oria I., I. Quicaño, E.Quispe and L. Alfonso, 2009. Variabilidad del color de la fibra de alpaca en la zona altoandina de Huancavelica-Perú (Variability of alpaca fibre colour in the highland region of Huancavelica-Peru). Animal Genetic Resources Information, 45: 79-84.

Quispe E.C., L. Alfonso, A. Flores, H. Guillén and Y. Ramos, 2009. Bases para establecer un programa de mejora de alpacas en la region altoandina de Huancavelica-Perú (Bases to an improvement program of the alpacas in highland region at Huancavelica-Peru). Archivos de Zootecnia, 58 (224): 705-716.

A microsatellite study on the genetic distance between Suri and Huacaya phenotypes in Peruvian alpaca (*Vicugna pacos*)

V. La Manna[1], A. La Terza[1], S. Dharaneedharan[1], S. Ghezzi[3], S. Arumugam Saravanaperumal[1], N. Apaza[4], T. Huanca[4], R. Bozzi[3] and C. Renieri[1]
[1]*Department of Environmental Sciences, University of Camerino, Italy;*
vincenzo.lamanna@unicam.it
[2]*INIA, ILLPA Puno, Rinconada Salcedo, Puno, Peru*
[3]*Department of Agricultural Biotechnologies, University of Florence, Italy*

Abstract

Two coat phenotypes exist in alpaca: the Huacaya and the Suri. The two coats have different textile characteristics and different prices on the market. Although present scientific knowledge suggests a simple genetic model of inheritance, there is a tendency to manage and consider the two phenotypes as two different breeds. A 14-microsatellite panel was used in this study to assess genetic distance between Suri and Huacaya alpacas in a sample of non-related animals from two phenotypically pure flocks at the Illpa-Puno experimental station in Quimsachata, Peru. The animals are part of a germplasm established approximately 20 years ago and have been bred separately according to their coat type since then. Genetic variability parameters were also calculated. The codominant data was statistically analysed using the software Genalex 6.3, Phylip 3.69 and Fstat 2.9.3.2. The sample was tested for Hardy-Weinberg equilibrium (HWE) and after strict Bonferroni correction only one locus (LCA37) showed deviation from equilibrium ($P<0.05$). Linkage disequilibrium (LD) was also tested and 9 loci associations showed significant disequilibrium. Observed heterozygosis (Ho= 0.766; SE=0.044), expected heterozygosis (He=0.769; SE=0.033), number of alleles (Na=9.667, SE=0.772) and Fixation index (F=0.004; SE=0.036) are comparable to data from previous studies. Measures of genetic distance were 0.06 for Nei's and 0.03 for Cavalli-Sforza's. The analysis of molecular variance reported no existing variance between populations. Considering the origin of the animals, their post domestication evolution and the reproductive practices in place, the results suggest that there is no genetic differentiation between the two populations for the studied loci.

Keywords: alpaca, *Vicugna pacos*, microsatellite, genetic distance, genetic diversity

Introduction

Two coat phenotypes exist in alpaca, namely the Huacaya and the Suri. They differ phenotypically in terms of fleece structure, the first one being a classically built fleece, compact, bulky and with high fibre crimp, similar to a merino fleece and the second one a fleece with longer and lustrous fibre organised in defined, 'hanging', locks, more similar to a Lincoln or to an Angora type of fleece. These two phenotypes also differ with regard to fibre structure and characteristics (Antonini *et al.*, 2001; McGregor and Butler, 2004; Frank *et al.*, 2006).

In terms of demography, there is a predominance of the Huacaya type which represents 90% of the alpaca fleece processed in Peru. This occurs despite the fact that the Suri trait seems to segregate in a very similar way to a single dominant gene or a haplotype (Ponzoni *et al.*, 1997). A different distribution of phenotypes can be observed in North America, Europe and Australia, where growing interest from both the textile industry and breeder associations in the Suri fibre has led to increasing Suri/Huacaya ratios and often to a paid premium for Suri fibre in the same fineness range as Huacaya (McGregor, 2006). Over the years this trend has generated two different schools of thought about the way the two phenotypes should be bred, classified and genetically managed. This is mainly true at a farming and breeder association level in North America and in Australia where it is often recommended to consider and manage the two phenotypes as two separate breeds in order to keep the Suri line 'pure' (Baychelier, 2002). This is contrary to present scientific theories which suggest that the two phenotypes are qualitative traits determined by 1 or 2 loci (Presciuttini *et al.*, 2010).

Population genetics by means of microsatellite analysis

A 10 microsatellite markers panel has been used by the ARI (Alpaca Registry Inc.) since 1998, mainly for parentage verification, and has found other applications throughout the years in other fields such as genome mapping, population structure and comparable genome analysis. As highlighted by several authors (Munyard *et al.*, 2009; Qiu *et al.*, 2006) the continuous and ongoing effort in sequencing the alpaca genome will rapidly lead to a much larger set of markers as demonstrated by a number of studies published in the past decade (Reed and Chaves, 2008; Sarno *et al.*, 2000; Penedo *et al.*, 1999a,b). Among the authors mentioned above, Reed and Chaves (2008) report an additional 1,516 potential loci by blasting *bos taurus* SSRs and Munyard *et al.* (2009) have recently found a set of 9 tetranucleotide markers. Some of these markers have already been used to calculate genetic distances among different species of South American Camelids (Bustamante *et al.*, 2002; Wheeler *et al.*, 2003, 2006).

Given the peculiar post-domestication evolutionary history of the species and the increasing interest in Suri fibre, the aim of this study is therefore to use a microsatellite panel to study the genetic distance between Suri and Huacaya alpacas and to assess the amplitude of genetic variability in the Peruvian alpaca population. The studied population belongs to a germplasm founded about 20 years ago in Quimsachata, Puno province, Peru.

Materials and methods

Alpaca blood samples were collected in spring 2008. Animals were kept and managed in the Illpa-Puno Experimental Station in Quimsachata (Puno province, Peru) at an altitude of approx. 4,200 m above sea level. The animals belonged to two geographically separate and phenotypically pure flocks (one Suri and one Huacaya) that have been managed and bred separately since the alpaca germplasm was created 20 years ago. Blood samples were taken from 65 non-related animals which included all the available males (n=15) and a subset of females (n=50), in total 32 Huacaya (7 males and 25 females) and 33 Suri (8 males and 25 females). Due to the lack of basic facilities, blood was collected by spotting a total of 100 μl of blood on Whatman FTA Nucleic Acid Collection cards (# WB120205). Possibly due to the non-sterile and difficult conditions in facilities on the Peruvian plateau, not all samples allowed for a sufficient quantity of DNA and in order to minimise the number of missing data, the dataset was rearranged to obtain a final number of 49 individuals, 10 males (5 Huacaya and 5 Suri) and 39 females (19 Huacaya and 20 Suri) and 13 microsatellites (LCA 19 was not included in the analysis). All samples were processed in Italy and genomic amplification was carried out by LGS genetic laboratories (Cremona, Italy). The markers, dyes utilised, primer sequences and allele sizes for the 14 microsatellites are shown in Table 1. The panel has been optimised to be amplified in two multiplex reactions of 7 primer pairs each.

Table 1. Markers and multiplex reaction data. List of markers, respective primer pairs, alleles fragment lengths and dyes used in the multiplex reactions.

Marker	Fragment length	Alleles	Dye	5'-3' Fw Primer sequence	5'-3' Rev Primer sequence
LCA 19	80-122	17	Vic	TAAGTCCAGCCCCACACTCA	GGTGAAGGGGCTTGATCTTC
LCA 94	187-213	9	Pet	GTCCATTCATCCAGCACAGG	ACATTTGGCAATCTCTGGAGAA
YWLL 44	84-136	18	Ned	CTCAACAATGCTAGACCTTGG	GAGAACACAGGCTGGTGAATA
YWLL 36	136-176	17	Vic	AGTCTTGGTGTGGTGGTAGAA	TGCCAGGATACTGACAGTGAT
YWLL 43	128-164	10	Pet	ATACCTCTCTTGCTCTCTC	CCTCTACAACCATGTTAGCCA
YWLL 29	210-232	9	Fam	GAAGGCAGGAGAGAAAGGTAG	CAGAGGCTTAATAACTTGCAG
LCA 37	124-174	19	Fam	AAACCTAATTACCTCCCCCA	CCATGTAGTTGCAGGACACG
LCA 5	178-218	13	Vic	GTGGTTTTTGCCCAAGCTC	ACCTCCAGTCTGGGGATTTC
LCA 8	211-261	14	Pet	GCTGAACCACAATGCAAAGA	AATGCAGATGTGCCTCAGTT
LCA 65	159-193	14	Fam	TTTTTCCCCTGTGGTTGAAT	AACTCAGCTGTTGTCAGGGG
LCA 66	216-266	24	Ned	GTGCAGCGTCCAAATAGTCA	CCAGCATCGTCCAGTATTCA
YWLL 40	176-190	7	Ned	CACATGACCATGTCCCCTTAT	CCAGTGACAGTGTGACTAAGA
LCA 99	263-297	11	Vic	CAGGTATCAGGAGACGGGCT	AGCATTTATCAAGGAACACCAGC
YWLL 46	87-115	5	Fam	AAGCAGAGTGATTTAACCGTG	GGATGACTAAGACTGCTCTGA

Software and statistical analysis

The statistical analysis of the microsatellite data for the genetic variability measures, including the analysis of molecular variance (AMOVA), was performed using the latest version of the software Genalex 6.3 (Peakall and Smouse, 2006), while the excel microsatellite toolkit (Park, 2001) was used for calculating the polymorphism information content (PIC) for each allele. The statistical population genetics package Fstat 2.9.3.2 (Goudet, 1995) was used to calculate deviation from Hardy-Weinberg equilibrium (HWE) and genotypic disequilibrium among loci applying a strict Bonferroni correction for multiple comparisons. The test for HWE and the test for linkage disequilibrium were carried out using 1,300 randomisations (table wide level of significance at 1%). Finally, Cavalli-Sforza's chord distance and Reynolds-Weir Cockerham distance were calculated using Gendist, an application of the software package Phylip version 3.69 (Felsenstein, 1989).

Results

Hardy Weinberg equilibrium and linkage disequilibrium: a number of indices have been calculated only on 12 microsatellites, excluding YWLL43, which is linked to the X sex chromosome. Only one locus (LCA37) was found not to be in Hardy Weinberg equilibrium after strict Bonferroni correction, showing excess of homozygosis ($P<0.05$). When the sample was split into two different populations the result did not differ. The test for genotypic disequilibrium between pairs of loci showed 9 loci associations out of 65 to be in some degree of linkage disequilibrium. The loci LCA8, LCA66 and LCA65 appear in 8 of the 9 associations showing linkage disequilibrium.

Heterozygosis, polymorphism information index and Fixation index: when the dataset was considered as a single population, the average Ho for the 12 markers was high (Ho=0.766; SE=0.044), extremely close to the average He (He=0.769; SE=0.033), and the unbiased expected heterozygosis (UHe=0.778; SE=0.033), with an overall average fixation index of 0.004 (SE 0.036). The high mean number of alleles (Na=9.667; SE=0.77), effective alleles (Ne=4.89; SE=0.39) and the low fixation indices (F) confirm such high values of heterozygosis and genetic variability.

The polymorphism information content (PIC) for each locus is in line with previous findings and ranges from 0.411 for locus YWLL46 to 0.826 for locus YWLL44. Ho, He, UHe, PIC and F are shown in Table 2 for each locus and as a mean for all loci with the relative standard errors. When the sample was analysed as two separate populations, results did not differ significantly. The null hypothesis was tested for Ho ($P=0.69$), He ($P=0.61$), F ($P=0.95$), Ne ($P=0.69$), PIC ($P=0.61$) and Na ($P=0.37$). Genetic variability parameters for the two populations are listed in Table 3.

Genetic distance and AMOVA

Genetic distance calculated by Pairwise Population Matrix of Nei's Genetic Distance and Unbiased Nei's Genetic Distance were 0.062 and <0.0001, respectively. When calculated as the Cavalli-Sforza's chord distance, the result was 0.03. Reynolds - Weir Cockerham distance, suggested to be more precise in the calculation of genetic distances between closely related species and breeds (Laval *et al.*, 2002), resulted in a value of 0.04. The principal coordinates analysis (PCA) calculated on the distance matrix among the individuals in Figure 1, graphically shows how samples from the Suri and Huacaya data sets overlap and do not segregate into different groups. The first 3 dimensions of the PCA explain 64.64% of the total variance.

Table 2. Population genetics parameters for a single population

Locus	N	Na	Ne	Ho	He	UHe	F	PIC
YWLL46	45	6 (5)	1.753	0.444	0.430	0.434	-0.034	0.411
LCA65	44	13 (14)	5.500	0.909	0.818	0.828	-0.111	0.795
YWLL40	46	6 (7)	4.804	0.826	0.792	0.801	-0.043	0.759
LCA5	49	7 (13)	3.904	0.837	0.744	0.752	-0.125	0.702
LCA66	48	13 (24)	6.103	0.896	0.836	0.845	-0.071	0.816
LCA8	49	9 (14)	6.188	0.837	0.838	0.847	0.002	0.818
LCA99	48	11 (11)	4.031	0.646	0.752	0.760	0.141	0.728
YWLL44	49	11 (18)	6.411	0.776	0.844	0.853	0.081	0.826
LCA37	47	13 (19)	4.499	0.553	0.778	0.786	0.289	0.761
LCA94	47	7 (9)	4.098	0.681	0.756	0.764	0.099	0.727
YWLL36	49	10 (17)	6.164	0.918	0.838	0.846	-0.096	0.817
YWLL29	48	10 (9)	5.219	0.875	0.808	0.817	-0.082	0.786
YWLL43[1]	46	6 (10)	2.608	-	-	-	-	0.561
Mean	47.4	9.667	4.89	0.766	0.769	0.778	0.004	0.746
SE	0.48	0.772	0.388	0.044	0.033	0.033	0.036	0.033

N: number of individuals; Na: number of different alleles (in brackets values from previous studies); Ne: number of Effective alleles; Ho: Observed Heterozygosity; He: Expected Heterozygosity; UHe: Unbiased Expected Heterozygosity = (2N/(2N-1))×He; F: Fixation Index; PIC: Polymorphism information content.
[1] Ho, He, UHe and F not shown for X-linked YWLL43 locus.

All the variance observed in the two populations with the AMOVA test was due to variation within populations (100%) and not to variation between populations (0%). Genetic differentiation by AMOVA test was calculated both by Fst (0.0002) and Rst (0.012).

Discussion

In terms of genetic variability within the Peruvian alpaca sample analysed, all parameters and findings, such as number of alleles (Na=9.667; SE=0.772), number of effective alleles (Ne=4.89; SE=0.388), observed and expected heterozygosis (Ho=0.766, SE=0.044; He=0.769, SE= 0.033), show that the Peruvian population is still conserving high genetic variability and does not show any sign of artificial selection pressure for the studied loci; these should be neutral to selective pressure but have not been mapped at present. The low fixation indices for these loci confirm this interpretation of the data (F=0.004; SE=0.036) and suggest that the microsatellite panel is suitable for genetic diversity studies. Furthermore, the sample size, although small, is likely to be representative of the genetic diversity within this population. In fact, in comparison with the data collected and analysed by the ARI since 1998 and by several other authors (Reed and Chaves, 2008; Sarno *et al.*, 2000; Penedo *et al.*, 1998, 1999a,b; McPartlan *et al.*, 1998), the sample in this study has shown the presence of a high number of alleles, matching the whole allelic range for most of the loci and in two cases (YWLL29 and YWLL46) showing one extra allele when compared with previous publications. Values for the polymorphism information content (PIC=0.746, SE=0.033) were also in line with previous findings. Only one locus (YWLL46) showed a PIC value <0.7 (PIC=0.411), which reflects the lower than average Na and Ne found for this specific locus (Na=6; Ne=1.753). Nevertheless, this value is larger than that from previous bibliographic data (Lang *et al.*, 1996). When these parameters were evaluated separately in the two phenotypic groups they did not differ significantly.

Table 3. Population genetics parameters for the two populations

Pop	Locus	N	Na	Ne	Ho	He	UHe	F	PIC
Huac	YWLL46	21	5	1.42	0.33	0.29	0.30	0.12	0.28
	LCA65	21	9	4.34	0.95	0.77	0.78	0.23	0.73
	YWLL40	21	6	4.48	0.86	0.77	0.79	0.10	0.74
	LCA5	24	6	4.08	0.92	0.75	0.77	0.21	0.71
	LCA66	23	11	5.69	0.91	0.82	0.84	0.10	0.80
	LCA8	24	8	6.36	0.83	0.84	0.86	0.01	0.82
	LCA99	23	10	4.15	0.65	0.75	0.77	0.14	0.73
	YWLL44	24	9	5.79	0.71	0.82	0.84	0.14	0.80
	LCA37	24	9	4.16	0.50	0.76	0.77	0.34	0.73
	LCA94	24	6	3.61	0.62	0.72	0.73	0.13	0.69
	YWLL36	24	9	6.29	0.92	0.84	0.85	0.09	0.82
	YWLL29	23	9	5.24	0.82	0.80	0.82	0.02	0.78
	YWLL43§	22	5	2.28	-	-	-	-	0.51
	Mean	23	8.08	4.63	0.75	0.74	0.76	0.01	0.72
	SE	0.37	0.54	0.39	0.05	0.04	0.04	0.05	0.04
Suri	YWLL46	24	5	2.11	0.54	0.52	0.53	0.02	0.50
	LCA65	23	11	5.72	0.87	0.82	0.84	0.05	0.80
	YWLL40	25	6	4.92	0.80	0.79	0.81	0.00	0.76
	LCA5	25	6	3.38	0.76	0.70	0.71	0.08	0.65
	LCA66	25	9	6.28	0.88	0.84	0.85	0.04	0.82
	LCA8	25	9	5.68	0.84	0.82	0.84	0.01	0.80
	LCA99	25	8	3.86	0.64	0.74	0.75	0.13	0.71
	YWLL44	25	10	6.07	0.84	0.83	0.85	0.00	0.81
	LCA37	23	13	4.77	0.61	0.79	0.80	0.23	0.77
	LCA94	23	7	4.41	0.74	0.77	0.79	0.04	0.74
	YWLL36	25	8	5.99	0.92	0.83	0.85	0.10	0.81
	YWLL29	25	10	5.04	0.92	0.80	0.81	0.14	0.77
	YWLL43[1]	24	5	2.76	-	-	-	-	0.57
	Mean	24.4	8.50	4.85	0.78	0.77	0.79	0.01	0.74
	SE	0.26	0.67	0.36	0.03	0.02	0.02	0.03	0.02

N: number of individuals; Na: number of different alleles; Ne: number of Effective alleles; Ho: Observed Heterozygosity; He: Expected Heterozygosity; UHe: Unbiased Expected Heterozygosity = (2N/(2N-1)) × He; F: Fixation Index; PIC: Polymorphism information content
[1] Ho, He, UHe and F not shown for X-linked YWLL43 locus.

In terms of genetic distance and differentiation between the two phenotypes both the PCA analysis and the hierarchical clustering analysis show no separate segregation or grouping of Suri and Huacaya individuals.

When the analysis of molecular variance was carried out considering the two phenotypes as two separate populations, it clearly identified the source of all variance in the component 'within populations', excluding any source of variance to be found between populations. This result is supported by the Nei's index of genetic distance, Cavalli-Sforza's chord distance and Reynolds - Weir Cockerham distance which also show no differentiation between the two populations. The first measure assumes a stepwise mutation model in an infinite allele model with equilibrium

Principal coordinates

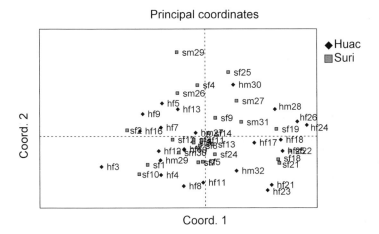

Figure 1. Principal Coordinates Analysis of Suri and Huacaya based on the distance matrix from molecular data.

between mutation and genetic drift, whereas the second and the third measures are dimensional models assuming only genetic drift.

There are two important factors to be taken into consideration while interpreting these results. The first is that the germplasm established 20 years ago at the experimental station of Quimsachata is not undergoing genetic selection and was created for the sole purpose of conserving the genetic diversity of the species. The second factor is that although Suri and Huacaya alpacas at the experimental station have been bred and managed separately since the creation of the germplasm, the time interval of 20 years is unlikely to generate genetic differentiation between the two phenotypes, especially considering the absence of selection, the reproductive physiology of the species and its generation time (Mason, 1973).

Conclusions

In light of these considerations, if a secondary breed structure had been present within the species at the time of the creation of the germplasm, it would have been preserved by the breeding practices in place at the experimental station. Nevertheless, the data obtained from the 13 loci suggest no genetic divergence between the two phenotypes and do not support the idea of two distinct populations of Peruvian Suri and Huacaya alpacas. Furthermore, the two phenotypes have similar genetic parameters in terms of allelic frequencies and genetic variability, showing high values both in terms of allelic richness and heterozygosis.

Acknowledgements

The authors would like to thank the LGS laboratories for carrying out the multiplex amplifications and band scoring.

References

Antonini M., F. Perdomici, S. Catalano, E.N. Frank, M. Gonzales, M.V.H. Hick and F. Castrignanò, 2001. Cuticular cell mean scale frequency in different types of fleece of domestic South American camelids (SAC). In: M. Gerken and C. Renieri (eds.), Progress in South American Camelids Research, Proceedings of the 3rd European Symposium and SUPREME European Seminar, 27-29 May 1999, 2001, EAAP110-116.

Baychelier P., 2002. What is a pure Suri? Alpacas Australia, 39: 30-33.

Bustamante A.V., A. Zambelli, D.A. De Lamo, J. von Thungen and L. Vidal-Rioja, 2002. Genetic variability of guanaco and llama populations in Argentina. Small Ruminant Research, 44: 97-101.

Felsenstein J., 1989. Notices; PHYLIP - Phylogeny Inference Package (Version 3.2). Cladistics, 5: 164-166.

Frank E.N., M.V.H. Hick, C.D. Gauna, H.E. Lamas, C. Renieri and M. Antonini, 2006. Phenotypic and genetic description of fibre traits in South American domestic camelids (llamas and alpacas). Small Ruminant Research, 61: 113-129.

Goudet J., 1995. FSTAT (Version 1.2): A Computer Program to Calculate F-Statistics. Journal of Heredity, 86: 485-486.

Lang K.D.M., Y. Wang and Y. Plante, 1996. Fifteen polymorphic dinucleotide microsatellites in llamas and alpacas. Animal Genetics, 27: 293.

Laval G., M. SanCristobal and C. Chevalet, 2002. Measuring genetic distances between breeds: use of some distances in various short term evolution models. Genetics, selection, evolution, 34: 481-507.

Mason I.L., 1973. The role of natural and artificial selection in the origin of breeds of farm animals. Zeitschrift für Tierzüchtung und Züchtungsbiologie, 90: 229-244.

McGregor B.A. and K.L. Butler, 2004. Sources of variation in fibre diameter attributes of Australian alpacas and implications for fleece evaluation and animal selection. Australian Journal of Agricultural Research, 55: 433-442.

McGregor B.A., 2006. Production, attributes and relative value of alpaca fleeces in southern Australia and implications for industry development. Small Ruminant Research, 61: 93-111.

McPartlan H.C., M.E. Matthews and N.A. Robinson, 1998. Alpaca microsatellites at the VIAS A1 and VIAS A2 loci. Animal Genetics, 29: 158-159.

Munyard K.A., J.M. Ledger, C.Y. Lee, C. Babra and D.M. Groth, 2009. Characterization and multiplex genotyping of alpaca tetranucleotide microsatellite markers. Small Ruminant Research, 85: 153-156.

Park S.D.E., 2001. Trypanotolerance in West African Cattle and the Population Genetic Effects of Selection. Thesis, University of Dublin, Dublin, Ireland.

Peakall R. and P.E. Smouse, 2006. Genalex 6: genetic analysis in Excel. Population genetic software for teaching and research. Molecular Ecology Notes, 6: 288-295.

Penedo M.C.T., A.R. Caetano and K.I. Cordova, 1998. Microsatellite markers for South American camelids. Animal Genetics, 29: 411-412.

Penedo M.C., A.R. Caetano and K. Cordova, 1999a. Eight microsatellite markers for South American camelids. Animal Genetics, 30: 166-167.

Penedo M.C., A.R. Caetano and K.I. Cordova, 1999b. Six microsatellite markers for South American camelids. Animal Genetics, 30: 399.

Ponzoni R.W., D.J. Hubbard, R.V. Kenyon, C.D. Tuckwell, B.A. McGregor, A. Howse, I. Carmichael and G.J. Judson, 1997. Phenotypes resulting from Huacaya by Huacaya, Suri by Huacaya and Suri by Suri crossings, In: Proceedings of the twelfth Conference of the Association for the Advancement of Animal Breeding and Genetics, 6-10 April 1997 1997, AAABG136-139.

Presciuttini S., A. Valbonesi, N. Apaza, M. Antonini, T. Huanca and C. Renieri, 2010. Fleece variation in alpaca (*Vicugna pacos*): a two-locus model for the Suri/Huacaya phenotype. BMC genetics, 11: 70.

Qiu J., T.E. Mitchell, A.P. Nickel, S.R. Brummet and W.E. Johnson, 2006. Identification and Characterization of Simple Sequence Repeat Markers in Alpaca (*Lama pacos*), In: Plant and Animal Genome XIV Conference Abstracts, 14-18 January 2006, National Center of Plant Gene Research (NCPGR)P134.

Reed K.M. and L.D. Chaves, 2008. Simple sequence repeats for genetic studies of alpaca. Animal Biotechnology, 19: 243-309.

Sarno R.J., V.A. David, W.L. Franklin, S.J. O'Brien and W.E. Johnson, 2000. Development of microsatellite markers in the guanaco, *Lama guanicoe*: utility for South American camelids. Molecular ecology, 9: 1922-1924.

Wheeler J.C., L. Chikhi and M.W. Bruford, 2006. Genetic Analysis of the Origins of Domestic South American Camelids. In: M.A. Zeder, D.G. Bradley, E. Emshwiller and B.D. Smith (eds.), Documenting domestication: new genetic and archaeological paradigms. University of California Press, Berkely, CA, USA, pp. 331-343.

Wheeler J.C., M. Fernandez, R. Rosadio, D. Hoces, M. Kadwell and M.W. Buford, 2003. Genetic diversity and management implications for vicuña populations in Peru. In: J. Lemons, R. Victor and D. Schaffer (eds.), Conserving biodiversity in arid regions: Best practices in developing nations. Kluwer Acadademic Publishers, Boston, MA, USA, pp. 327-344.

Genetic differentiation of six Peruvian alpaca populations

M. Paredes[1], J. Machaca[2], P.J. Azor[1], A. Alonso-Moraga[1], A. Membrillo[1] and A. Muñoz-Serrano[1]
[1]*Dpto. De Genética. Universidad de Córdoba. Campus de Rabanales C-5, 14071 Córdoba, Spain; mparedes@ufro.cl*
[2]*Centro de Estudios y Promoción del Desarrollo DESCOSUR. Málaga Grénet N° 678 Umacollo, Arequipa, Peru*

Abstract

We analysed 227 individuals belonging to six populations of alpaca from the high-Andean zones in Peru, where livestock production is based on the breeding of domestic South American camelids (alpacas and llamas), basically alpacas from the Huacaya breed. Twenty microsatellite markers were analysed and genetic variability and genetic differentiation parameters were estimated. Expected heterozygosity (He) ranged from 0.660 in population 5 to 0.717 in population 3. Observed heterozygosity (Ho) oscillated from 0.552 in population 2 to 0.637 in population 4. Average number of alleles per locus in each population ranged from 5.65 in population 2 to 9.2 in population 1. In order to analyse the genetic structure and genetic differentiation we estimated the Wright's F-statistics. F_{ST} values between populations ranged from 0.00466 between 4 and 6 populations pair and 0.4998 between 4 and 5 populations pairs. Average F_{ST} value between populations was 0.0299 (0.0212 - 0.0408). Average F_{IT} and F_{IS} values were 0.1742 (0.0789 - 0.2968) and 0.14874 (0.0551 - 0.2664) respectively. Estimated number of migrants per generation between populations was very high as we detected a scarce genetic differentiation between them. The high level of observed homozygosity may be due to inbreeding.

Keywords: animal genetic resources, biodiversity, conservation, genetic variability, sustainable development

Introduction

Animal production plays an important role in the quality of products, production under less intensive conditions, more respect for the environment, guarantees for the sustainability of the system and efforts to maintain the population in the rural areas. Within these strategies it is desirable on the one hand to maintain biodiversity, preventing the loss of genetic patrimony, and on the other hand, to preserve local breeds to help the subsistence of rural populations.

Peru is the world's number one producer of camelid fibre (90% of global production) accounting for more than three million alpacas. Livestock production in the high-Andean zones in Peru is based on the breeding of domestic South American camelids (alpacas and llamas), basically alpacas from the Huacaya breed.

This animal production is the main economic activity of the residents, focused on fibre production. So the genetic improvement of this breed in the high-Andean zones is very important in order to improve the fibre quality traits and consequently the quality of life of autochthonal breeders.

Molecular markers offer an unbiased system that allows for the genetic characterisation of populations of this species (Kadwell *et al.*, 2001) and the population structure analysis. Microsatellite markers are considered the best tools for genetic identification, molecular phylogeny and parentage testing in breeds (Aranguren-Mendez and Jordana, 2001; Schlötterer and Harr, 2001).

One of the main problems in the populations is the increase in the inbreeding level and consequently inbreeding depression: a reduction of average phenotypic values of the traits (Falconer and Mackay, 1996). When genealogical information is not reliable, molecular tools have to be implemented in order to estimate the level of genetic variability (Azor and Goyache, 2007).

Material and methods

We have analysed 227 individuals (males and females) belonging to six populations of alpaca from the high-Andean zones in Peru (Table 1). Samples were taken from ear and DNA was isolated using the i-genomic CTB® commercial kit.

Twenty microsatellite markers were amplified using multiplex PCR methods and analysed in all individuals sampled (Table 2).

Allele sizes were analysed in an ABI 3130 capillary automatic sequencer (applied biosystems). Genotyping was performed using GeneMapper 3.7 software. Expected (He) and observed heterozygosity (Ho) were estimated. F-Statistic (Wright, 1965): F_{IS}, F_{IT} y F_{ST} were estimated using the Weir and Cockerham (1984) methodology. The bootstrapped confidence interval was based on 1000 replications. Effective number of migrants per generation (Nm) was also estimated (Slatkin, 1985).

Results and discussion

All genetic markers used were polymorphic. Average number of allele per locus oscillated between 5.65 in population number 2 and 9.1 in population number 1 (Table 3). These values are similar to others found in other artiodactyls. Sastre *et al*. (2007) found in the Creole Casanare cattle breed an average 8.54 number of alleles per locus and 6.59 in zebu. Azor *et al*. (2004) found similar values in Merino breed and higher in Fleischschaf (5.03), Ile de France (5.63) or other endangered populations such as Merino Negro (5.64) and Churra Lebrijana (4,75) (Azor *et al*., 2004) and also in Murciano Granadina goat breed (7.3), Malagueña (7.4) or Creole breeds (Azor *et al*., 2008). In Iberian pig, Duroc and Alentejano breeds Membrillo *et al*. (2007) values were lower than in alpacas (2.3-4.8).

Observed heterozygosity (Ho) per population oscillated between 0.5524 in population number 2 and 0.6379 in population number 4 (Table 3). Azor *et al*. (2003) found a higher Ho value (0,685) in Pajuna cattle breed and Sastre *et al*. (2007) showed also higher values in zebu 0.669 and 0.727 in Creole Casanare cattle breed. In Merina and Montesina sheep breeds Azor *et al*. (2004) and Valle *et al*. (2004) showed higher values: 0.6967 and 0.6527 respectively. We

Table 1. Number of individuals sampled per population.

Population	Zone	Area	Region	Sample size
1	San Juan de Tarucani	San Juan de Tarucani	Arequipa	69
2	Estación Pillones	San Antonio de Chuca	Arequipa	24
3	Colca	San Antonio de Chuca	Arequipa	49
4	Chalhuanca	Yanque	Arequipa	17
5	Lampa	Lampa	Puno	36
6	Palca	Palca	Puno	32

Table 2. Microsatellite markers analysed in alpaca.

Marker	Primers 5'-3'(forward and reverse)	Size	Fluoro-chrome	Tᵃ	Reference
VOLP12	D-TTGTTCTCAACAGGGACTGC R-TCTGGCCCACCCACTAA	130-164	PET	55.2	Obreque et al., 1998a
CVRL06	D-TTTTAAAAATTCTGACCAGGAGTCTG R-CATAATAGCCAAAACATGGAAACAAC	190-240	PET		Mariasegarem et al., 2002
VOLP01	D-CCCATTGGCGATTATTTAGG R-AACAGGGGTAACAAACAGGAC	241-267	PET		Obreque et al., 1999b
LCA23	D-TCACGGCAAAGACGTGAATA R-CCAGGAAACACACACCCAC	120-160	6-FAM	55.2	Penedo et al., 1998a
LCA65	D-TTTTTCCCCTGTGGTTGAAT R-AACTCAGCTGTTGTCAGGGG	161-193	6-FAM		Penedo et al., 1998b
CVRL07	D-AATACCCTAGTTGAAGCTCTGTCCT R-GAGTGCCTTTATAAATATGGGTCTG	268-286	PET		Mariasegarem et al., 2002
VOLP04	D-GCATTTCTCCGTAATCATTG R-TGACACCTTTTGTTTCCATT	214-260	6-FAM	55.8	Obreque et al., 1999b
CVRL05	D-CCTTGGACCTCCTTGCTCTG R-GCCACTGGTCCCTGTCATT	109-151	VIC		Mariasegarem et al., 2002
CVRL01	D-GAAGAGGTTGGGGCACTAC R-CAGGCAGATATCCATTGAA	159-201	VIC		Mariasegarem et al., 2002
CVRL08	D-AATTCCTGTGATTTTATACACA R-CATGTCATGAAAGCTACAGTA	194-216	NED		Mariasegarem et al., 2002
LGU56	D-TTGCTGTACCGGAGATGTTG R-TTGAGGCAAGAATGCAGATG	157-183	6-FAM	55.2	Sarno et al., 2000
LGU68	D-CATCTACATGCCCCTGTGTG R-TGCAGGGAGGACTAACAGGT	202-232	6-FAM		Sarno et al., 2000
VOLP77	D-TATTTGGTGGTGACATT R-CATCACTGTACATATGAAGG	141-165	VIC		Obreque et al., 1999b
LGU50	D-CTGCTGTGCTTGTCACCCTA R-AGCACCACATGCCTCTAAGT	168-192	VIC	57.8	Sarno et al., 2000
LGU49	D-TCTAGGTCCATCCCTGTTGC R-GTGCTGGAATAGTGCCCAGT	218-244	VIC		Sarno et al., 2000
LCA71	D-CAGACATATACCTGTATCCGTATCTA R-TTCAGTGTTTCCTCGCAATG	130-150	NED		Penedo et al., 1998b
LAB1	D-AGAGGATCAATCCCTCTGAGAT R-ATTAGAGGCCAGTATAACAATC	154-180	NED	55.2	Bustamante et al., 2003
LGU51	D-CCTTCCTCTTGCAAATCTGG R-GCACCTGATGTCATTTATGAGG	192-208	NED		Sarno et al., 2000
YWLL08	D-ATCAAGTTTGAGGTGCTTTCC R-CCATGGCATTGTGTTGAAGAC	126-184	PET		Lang et al., 1996
LGU79	D-TAAGGTAGGAGCGAGCCAAA R-ACCTGCTCGCTAATCTCTGC	197-223	PET		Sarno et al., 2000

have obtained a lower *Ho* value in alpaca than that obtained in Murciano Granadina (0.6812), Malagueña (0.6452) and Creole Peruvian (0.7076) goat breeds, and similar to 0.5904 found in Creole Mexican and higher than 0.5382 in Creole Chilean (Azor *et al.*, 2008).

Breeding and genetics

Table 3. Expected and observed heterozygosity (He and Ho) values and average number of alleles per locus.

Population	He	Ho	Average number of alleles per locus
1	0.7095	0.5986	9.2000
2	0.6639	0.5524	5.6500
3	0.7179	0.6174	8.2500
4	0.6835	0.6379	6.7000
5	0.6602	0.5934	6.7000
6	0.6636	0.6035	7.1500

Expected heterozygosity (He) values oscillated between 0.66 in populations 2, 5 and 6 and 0.718 showed in population 3 (Table 3). *He* values have been lower than 0.7462 found in Merina sheep breed and higher than 0.5829 found in Fleischschaf sheep breed, 0.6120 in Ile de France, 0.6147 in Merino Negro, 0.6432 in Merino Preto portugués and 0.5804 in Churra Lebrijana (Azor *et al.*, 2004). Azor *et al.* (2008) found lower He values in Spanish and Creole goat breeds.

F_{ST} values between alpaca population pairs (Table 4) were very similar. They oscillated between 0.01143 (population 1-4) and 0.04998 (population 4-5). In general the highest values were found between population 5 and the others populations. The Nm values that provide information about the gene flow between populations were very high, especially between population pairs: 4-6, 1-3, 1-4 and 1-6. No relationship was found between the geographical vicinity and the level of gene flow.

Populations F-statistics values are shown in Table 5. F_{IS} value shows that there is a reduction in heterozygosity due to inbreeding in populations and F_{IT} value shows that there is a reduction in heterozygosity due to inbreeding in the total population. Fixation index or genetic differentiation coefficient (F_{ST}) confirms that populations are very similar and there is a very high gene flow.

Table 4. F_{ST} values between alpaca population pairs (above diagonal) and Nm values (below diagonal).

Population	1	2	3	4	5	6
1		0.02793	0.01575	0.01143	0.04149	0.01174
2	8.70		0.03288	0.03324	0.04941	0.02907
3	15.62	7.35		0.03016	0.04879	0.03469
4	21.62	7.27	8.04		0.04998	0.00466
5	5.78	4.81	4.87	4.75		0.04190
6	21.05	8.35	6.96	53.38	5.72	

Table 5. F-statistics values in six alpaca populations.

	F_{IS}	F_{IT}	F_{ST}
Between populations	0.1487	0.1742	0.0299
Confidence interval (95%)	(0.0551-0.2664)	(0.0790-0.2968)	(0.0212-0.0408)

Conclusions

We have detected a very high gene flow between populations and a scarce genetic differentiation between them. Therefore, the high number of alleles detected and the high values of He and Ho will permit us to design a promising breeding programme in this breed in order to improve the fibre quality traits and increase the farms' profitability.

Acknowledgements

This study was partially funded by a project from the Vicerrectorado de Internacionalización y Cooperación Internacional of University of Córdoba, Spain, entitled *Mejora de la capacidad productiva de fibra de alpaca (*Vicugna pacos*) en las comunidades alto andinas de la región Arequipa, Peru.*

References

Aranguren-Mendez J. and J. Jordana, 2001. Utilización de marcadores de ADN (microsatélites) en poblaciones de animales domésticos en peligro de extinción. Memorias Asociación Venezolana de Producción Animal (AVPA). Venezuela.

Azor P.J., A. Molina, M. Valera and A. Luque, 2003. Diversidad genética de las subpoblaciones de ganado bovino Pajuno. VIII Jornadas Científicas de Veterinaria Militar, Madrid, Spain.

Azor P.J., A. Molina, F. Barajas, J.J. Arranz, M. Valera, A. Rodero and J. Miguélez, 2004. Estimación del nivel de diferenciación genética de la raza merina mediante ADN Microsatélite, FEAGAS, 25: 92-98.

Azor P.J. and F. Goyache, 2007. Metodología de caracterización genética. In: Patrimonio Ganadero Andaluz. Volume I. La ganadería andaluza en el Siglo XXI., Junta de Andalucía (ed.). Sevilla, Spain, pp. 483-524.

Azor P.J., M. Valera, J. Sarria, J.P. Avilez, J. Nahed, M. Delgado and J.M. Castel, 2008. Estimation of genetic relationships between Spanish and Creole goat breeds using microsatellite markers. (Estimación de relaciones genéticas entre las razas caprinas Española y Creole usando marcadores microsatélites). ITEA, 104: 323-327.

Bustamante A.V., M.L. Maté, Zambelli and L. Vidal-Rioja, 2003. Isolation and characterization of 10 polymorphic dinucleotide microsatellite markers for llama and guanaco. (Aislamiento y caracterización de 10 marcadores microsatélites dinucleotídicos polimórficos para la llama y el guanaco). Molecular Ecology Notes, 3: 68-69.

Falconer D.S. and T. Mackay, 1996. Introduction to quantitative genetics. A.W. Longman Limited, Essex, UK.

Kadwell M., M. Fernandez, H.F. Stanley, R. Baldi, J.C. Wheeler, R. Rosadio and M.W. Brudford, 2001. Genetic analysis reveals the wild ancestors of the llama and the alpaca. Proceedings of the Royal Society of London. Series B, Biological Sciences, 268 (1485): 2575-2584.

Lang K., Y. Yang and Y. Plante, 1996. Fifteen polymorphic dinucleotide microsatellites in llamas and alpacas. Animal Genetics, 27: 285-294.

Mariasegaram M., S. Pullenayegum, M. Jahabar Ali, R.S. Shah, M.C.T. Penedo, U. Wernery and J. Sasse, 2002. Isolation and characerization of eight microsatellite markers in *Camelus dromedarius* and cross-species amplification in C. bactrianus and Lama pacos. Animal Genetics, 33: 377-405.

Membrillo A., P.J. Azor I. Clemente, G. Dorado, E. Diéguez, A. Jiménez, E. Santos and A. Molina, 2007. Estudio de las relaciones genéticas de las estirpes del cerdo ibérico mediante marcadores microsatélites. IV Jornadas Ibéricas de razas autóctonas y sus productos tradicionales: Innovación, seguridad y cultura alimentaria. Sevilla, Spain.

Obreque V., L. Coogle, P.J. Henney, E. Bailey, R. Mancilla, J. García-Hiudobro, P. Hinrichsen and E.G. Cothran, 1998a. Characterization of 10 polymorphic alpaca dinucleotide microsatellites. Animal Genetics, 29: 460-477.

Obreque V., R. Mancilla, J. García-Hiudobro, E.G. Cothran and P. Hinrichsen, 1999b. Thirteen new dinucleotide microsatellites in Alpaca. Animal Genetics, 30: 382-405.

Penedo M.C.T., A.R. Caetano, K.I. Cordova, 1998a. Microsatellite markers for South American camelids. (Marcadores microsatellite para Camélidos Sudamericanos). Animal Genetics, 29: 398-413.

Penedo M.C.T., A.R. Caetano and K. Cordova, 1998b. Eight microsatellite markers for South American camelids. Animal Genetics, 30: 161-168.

Breeding and genetics

Sarno R., V. David, W. Franklin, S. O'Brien and W. Johnson, 2000. Development of microsatellite markers in the guanaco, *Lama guanicoe*: utility for South American camelids. Molecular Ecololy, 9: 1919-1952.

Sastre H., E. Rodero, A. Rodero, P.J. Azor, N. Sepúlveda, M. Herrera and A. Molina, 2007. Genetic Study of the Colombian Criollo Casanare Breed and the Relationship With Others cattle Breeds (Estudio genético de la raza Casanare criolla colombiana y su relación con otras razas vacunas). Revista Científica, FCV-LUZ/Vol. XVII, N° 5, pp. 1-9.

Schlötterer C., B. Harr, 2001. Microsatellite Instability. Encyclopedia of life sciences. (Inestabilidad de microsatelites. Enciclopedia de las ciencias de la vida), Nature Publishing Group: 1-4.

Slatkin M. 1985. Gene flow in natural populations. Annual Review of Ecology Systematics, 16: 393-430.

Valle J.P., J. Azor, M. Valera, J.J. Arranz and A. Molina, 2004. Análisis de la variabilidad genética de la raza Montesina mediante marcadores de ADN. FEAGAS, 25: 99-105.

Weir B.S. and C.C. Cockerham, 1984. Estimating F-statistics for the analysis of population structure. Evolution, 38: 1358-70.

Wright S. 1965. The interpretation of population structure by F-statistics with special regard to systems of mating. Evolution, 19: 395-420.

Nutrition and reproduction

Lactation in llamas (*Lama glama*): estimating milk intake and output using stable isotope techniques

A. Riek[1,2] and M. Gerken[1]
[1]Department of Animal Sciences, University of Göttingen, Albrecht-Thaer-Weg 3, 37075 Göttingen, Germany; mgerken@gwdg.de
[2]Present address: Centre for Behavioural and Physiological Ecology, Zoology, University of New England, Armidale NSW 2351, Australia

Abstract

Compared to many other domesticated ruminants, very little is known about lactation in South American camelids. However, knowledge about milk and milk nutrient intakes in suckling young is essential for giving recommendations on an adequate nutrient supply. Direct measurement of milk production in South American camelids is hardly applicable, due to their short teats of about 2 cm and the low storage capacity of the udder. Therefore, we estimated milk production in llamas by measuring the milk intake in suckling young as an indirect trait. The study was conducted in Germany under temperate climatic conditions involving 11 lactating dams at two consecutive lactation periods (average body mass: 147.7±18.2, mean ± SD). Milk intake was estimated from water kinetics of 17 suckling young using two different stable isotope techniques, namely the isotope dilution technique involving the application of deuterium oxide (D_2O) to the suckling young and the 'dose-to-the-mother' technique involving the application of D_2O to the lactating dam. The latter method additionally allows for the estimation of non-milk water intakes of suckling young. Daily milk intakes averaged 2.7±0.63, 2.2±0.60 and 2.0±0.51 kg at 3-4, 10-11 and 18-19 weeks *post partum*. Milk intake in suckling young decreased with age when expressed as daily amount, percentage of body mass or per kg metabolic size ($P<0.001$), but the influence of age was eliminated when expressed per g daily gain. In suckling llamas, total water turnover increased with age ($P<0.01$), whereas the milk water fraction decreased ($P<0.001$). With increasing age less water from milk and more water from other sources was ingested. In contrast, the fraction of the milk water excreted by the dam did not change, indicating a fairly constant relationship between milk-water excreted via milk and total water ingested during lactation. Our results show that the stable isotope techniques used in the present study give reasonable estimates of the milk production in llamas. Furthermore, combined with milk composition data, the present milk intake estimations at different stages of lactation can be used to establish recommendations for nutrient and energy requirements of suckling llamas and lactating dams.

Keywords: llama, milk, lactation, deuterium, stable isotopes

Introduction

Despite the growing interest in llamas in Europe, North America and Australia as farm and pet animals, very little information is available on the milk output and the rearing of suckling llamas. Hand-milking is hardly applicable to llamas due to the short teats of about 2 cm (Fowler, 1989). In addition, the storage capacity of the udder is very limited and milking would need to take place every 2-3 hours according to the natural suckling interval of nursing llamas (Pouillon, 2001).

Stable isotope dilution techniques offer a viable and relatively precise method for measuring milk intake and output. One of these methods is the isotope dilution (ID) technique, which measures the decline of a known administered amount of hydrogen isotope in body fluids from the dilution by ingested water in the suckling young. Another method is the isotope transfer (IT) method, based on the transfer of one water isotope from the lactating mother to the suckling young via milk and the concurrent determination of the water turnover of the young by another hydrogen isotope. A detailed description of both techniques is given elsewhere (Holleman *et al.*, 1975, 1982; Oftedal, 1981; Dove, 1988). The IT technique has the advantage that no other sources of water intake, such as drinking water, water from feedstuff, etc., ingested by the suckling young need to be accounted for to estimate the milk intake. Furthermore, the water derived from milk and from other sources can be distinguished, thus allowing the estimation of total milk intake and water intake rate. However, as regulations for the use of radioactive substances and their disposal become stricter, the experimental application of ^{3}H is increasingly difficult, especially in larger animals.

An alternative approach is the so-called 'dose-to-the-mother' (DTTM) technique (Coward *et al.*, 1979), which involves the application of D_2O to a lactating female. By measuring the transfer of D_2O from the mother to the suckling young via milk, total daily milk intake can be calculated. This technique has so far been applied mainly to humans (Coward *et al.*, 1979; Butte *et al.*, 1988; Haisma *et al.*, 2003). The advantage of this technique over the ID technique is that by applying some assumptions it allows the estimation of other sources of water intake in the young than water intake via milk.

Therefore, the present investigation summarises published milk intake and output data in llamas measured by the ID and DTTM method. Possible recommendations for the energy supply in suckling llamas were deduced. Furthermore, calculated data on the total water turnover (TWT) and water intake in the lactating dam, allowed the estimation of water consumption during lactation in the dam.

Materials and methods

Animals and management

In three trials, a total of 11 female llamas and 17 nursing young (crias) were involved, originating from a herd of the Experimental Station Relliehausen of Göttingen University and a private German breeder. Animals were transferred 3 months prior to parturition for acclimatisation and were kept at the Department of Animal Sciences, University of Göttingen, Germany, for a lactation period of 27 weeks under controlled stable conditions. Each room measured 5.8 by 3.2 m and animals had permanent access to an outdoor pen. In the stable, light schedule was kept constant (14 h light to 10 h dark). Llama dams were fed twice daily 0.5 kg of a commercial mixed grain and molasses feed containing 16.0% CP, 12.0% crude fibre, 3.0% crude fat, 1.2% calcium 0.5% phosphorus, 0.3% sodium, 8.5% ash and 10.2 MJ/kg of ME (HG 58 S, Raiffeisen-AGRAVIS AG, Rosdorf, Germany). Hay from ryegrass dominated grassland (DM: 860 g/kg fresh matter, crude ash: 94 g/kg DM, CP: 131 g/kg DM, ether extract: 26 g/kg DM, crude fibre: 283 g/kg DM, nitrogen free extractives: 466 g/kg DM), water and mineral feed (HG MIN 13, Raiffeisen-AGRAVIS AG, Minden Westf., Germany) were available *ad libitum*.

Crias had access to hay, as hay was considered a negligible source of extraneous water for the measurement periods. Water intake by the crias was restricted to dam's milk until week 10 *post partum* (PP), since in a pre-trial, crias did not start to consume drinking water until week 11 PP. After that, crias had access to water *ad libitum*, except for the measurement periods when water intake was restricted to milk water.

Milk intake studies

Milk intake was estimated from water kinetics of 17 suckling young and 6 dams using two different stable isotope techniques, namely the isotope dilution (ID) technique and the 'dose-to-the-mother' (DTTM) technique (Coward *et al.*, 1979; Haisma *et al.*, 2003). For both techniques, the 3 measurement periods were at 3 to 4, 10 to 11 and 18 to 19 weeks PP, lasting 7 d each. Averages were calculated for the midpoint of each study period. In 11 crias milk intake was determined from water kinetics by calculating the wash-out rates of administered deuterium (D_2O). In another 6 crias milk intake was determined by the DTTM technique. The method depends on measuring the transfer of ^2H labelled water from the mother to the young via milk. A detailed description of both methods is given in Riek *et al.* (2007) and Riek and Gerken (2009).

Statistical analyses

Statistical analyses were performed with the software package Statistical Analysis Systems version 9.01 (SAS, 2001). Sex of cria exerted no significant influence in any of the trials. Accordingly, a two-way ANOVA was performed including only the effects of age and animal on various parameters, using the General Linear Model procedure (PROC GLM), with animal as random and age as fixed effect. An integrated multiple range test (Student-Newman-Keuls) was used to detect differences between means with a 5% significance level.

Results

Crias

In Table 1 measured and calculated data on body mass, daily mass gain, body water, milk intake, non-milk water intake and milk nutrients intake are presented. Body mass increased significantly with age ($P<0.001$), while daily mass gain revealed only a decreasing tendency. The water fraction, which is the relation between body water and body mass, decreased ($P<0.001$). By contrast, total water turnover (TWT) increased significantly with age, whether expressed as kg/d or per kg BW ($P<0.001$). Milk intake, calculated as daily amount, percentage of body mass or per kg metabolic mass, decreased significantly with age ($P<0.001$). However, when expressed as per g growth, the age effect was eliminated ($P=0.845$). Combining data on milk-water intake and TWT revealed a steady decline in the milk water fraction ($P<0.001$), indicating that with increasing age less water from milk and more water from other sources was ingested. Accordingly, non-milk water intake increased with age ($P<0.001$). Milk nutrient intakes, calculated by combining milk intake with the respective milk composition data from Table 2 decreased significantly with age for all nutrients.

Dams

Body mass of the dams changed significantly ($P<0.05$) between, but not during, measurement periods (Table 2). Body water and body water fraction ($P=0.072$ and $P=0.111$, respectively) were not affected by the lactation stage. The fractional water turnover rate instead steadily decreased with prolonged lactation ($P<0.01$). Accordingly, TWT declined whether expressed as daily amount, percentage of body mass, or per kg metabolic mass. However, the milk-water fraction did not change significantly ($P=0.398$), indicating a fairly constant relationship between milk-water secretion and total water ingested during lactation. Fat, DM and energy concentration of the milk increased significantly as lactation proceeded, while protein and lactose remained fairly constant.

Table 1. Milk, milk nutrient and non-milk water intakes of suckling llamas measured by the isotope dilution and the 'dose-to-the-mother' technique (see text for details; values are means ± SE) (modified after Riek et al., 2007; Riek and Gerken, 2009).

Parameter	n	Time post partum (d)			P-values	
		25	73	129	Age effect	Animal effect
Body mass (kg)	17	20.82±1.20[a]	32.67±2.18[b]	49.01±3.14[c]	< 0.001	0.003
Massgain (g/d)	17	360±32	313±27	263±20	0.413	0.315
Body water fraction (%)[1]	17	72.0±0.3[a]	70.4±0.5[b]	69.1±0.4[b]	< 0.001	< 0.001
Milk water intake (kg/d)	17	2.50±0.20[a]	2.07±0.25[b]	1.85±0.16[c]	< 0.001	< 0.001
Milk intake (kg/d)	17	2.67±0.14[a]	2.24±0.19[b]	1.97±0.31[b]	< 0.001	< 0.001
per cent of body mass (%)	17	12.8±0.4[a]	6.8±0.3[b]	4.0±0.4[c]	< 0.001	0.081
per kg (body mass)$^{0.75}$ (g)	17	274±12[a]	162±10[b]	106±14[c]	< 0.001	0.011
per g growth (g)	17	7.5±0.4	7.2±0.3	7.5±0.9	0.845	0.671
Total water turnover (kg/d)	6	2.87±0.20[a]	2.60±0.31[a]	3.92±0.51[b]	0.003	0.007
per kg (body mass) (%)	6	13.21±0.76[a]	7.25±0.60[b]	7.68±0.60[b]	< 0.001	0.366
per kg body water$^{0.82}$ (ml)	6	299±17[a]	185±17[b]	213±19[b]	0.002	0.155
Non-milk water intake (kg/d)	6	0.25±0.06[a]	0.54±0.11[a]	2.01±0.17[b]	< 0.001	0.499
Milk-water fraction (%)[2]	6	90.2±2.2[a]	78.5±3.9[b]	47.1±3.5[c]	< 0.001	0.344
Milk nutrient intakes	17					
Fat (g/d)	17	111±8[a]	100±11[ab]	93±9[b]	< 0.05	< 0.05
Protein (g/d)	17	108±8[a]	91±10[ab]	83±7[b]	< 0.01	< 0.001
Lactose (g/d)	17	159±13[a]	137±13[b]	116±10[b]	< 0.001	< 0.001
Dry matter (g/d)	17	404±34[a]	345±34[b]	306±27[b]	< 0.001	< 0.001
Gross energy (MJ/d)	17	9.68±0.73[a]	8.46±0.85[b]	7.61±0.69[c]	< 0.01	< 0.001

[a, b, c] Mean values within a row with different superscript letters were significantly different (P<0.05) based on Student-Newman-Keuls multiple range test for multiple comparisons.
[1] Relationship between body water and body mass.
[2] Relationship between milk-water and total water turnover.

Discussion

Crias

Llama crias consumed 764 kJ ME per kg BW$^{0.82}$ or 849 kJ GE per kg BW$^{0.82}$ at peak lactation, which is close to the value of 942 kJ GE per kg BW$^{0.83}$ derived by Oftedal (1981) from milk intake data of several species at peak lactation, including domestic and wild ruminants. Only two studies determined the maintenance requirements for llamas in adult animals (Schneider et al., 1974; Carmean et al., 1992). Based on these results, Van Saun (2006) suggests an average value for maintenance of 305 kJ ME/kg BW$^{0.75}$ per day for llamas. However, these values are not applicable for crias, whose only nutrient source is milk, because the respective variables need to be scaled to metabolic size raised to the power of 0.82 (Riek, 2008). As shown in Figure 1, predicted maintenance at zero growth for suckling llamas at peak lactation is 312 kJ/kg BW$^{0.82}$ per day corresponding to 3.69 MJ ME. The remaining ME from the energy consumed (5.38 MJ ME/d) is then assumed to be used for growth and corresponds to 15 kJ ME/g mass gain. This estimate is the same as reported by Johnson (1994) for suckling llamas, but no methodology was described.

Table 2. Body water turnover and milk composition in llama dams (numbers are means ± SE; n=6) (after Riek and Gerken, 2009).

Parameter	Time post partum (d)			P-values	
	25	73	129	Age effect	Animal effect
Body mass (kg)	148.7±11.4[a]	140.3±11.1[b]	139.3±9.9[b]	< 0.05	< 0.001
Fractional turnover (d[-1])	0.120±0.010[a]	0.105±0.008[a]	0.081±0.005[b]	< 0.01	< 0.05
Body water (kg)	93.2±5.8	86.5±6.7	91.3±6.4	0.072	< 0.001
Body water fraction (%)[1]	63.1±1.2	61.9±1.9	65.6±1.0	0.111	0.099
Total water turnover (kg/d)	10.9±0.98[a]	9.1±0.96[b]	7.3±0.71[c]	< 0.01	< 0.01
per kg (body mass)$^{0.75}$ (g)	263±23[a]	228±18[a]	180±11[b]	< 0.01	0.052
per cent of body mass (%)	7.6±0.7[a]	6.7±0.6[a]	5.3±0.3[b]	< 0.01	< 0.05
Milk-water fraction (%)[2]	24.8±2.9	23.2±2.4	23.7±2.4	0.398	< 0.001
Milk output (g/kg BW$^{0.75}$)	66.4±4.1[a]	54.3±5.2[b]	50.5±6.5[b]	< 0.01	< 0.001
Milk composition					
Fat (%)	3.97±0.20[a]	4.45±0.31[ab]	4.93±0.26[b]	< 0.01	< 0.05
Protein (%)	3.92±0.15	3.94±0.11	4.23±0.12	0.215	0.498
Lactose (%)	5.95±0.07	6.07±0.09	6.05±0.09	0.153	< 0.001
Dry matter (%)	14.43±0.24[a]	15.25±0.31[b]	15.97±0.33[c]	< 0.001	< 0.05
Energy (MJ/kg)	3.51±0.08[a]	3.72±0.12[b]	3.98±0.11[c]	< 0.001	< 0.01

[a, b, c] Mean values within a row with different superscript letters were significantly different (P<0.05) based on Student-Newman-Keuls multiple range test for multiple comparisons.
[1] Relationship between body water and body mass.
[2] Relationship between milk-water excreted via milk and total water turnover.

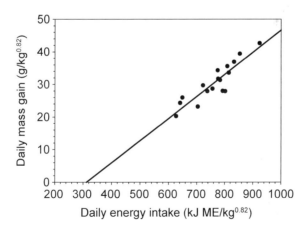

Figure 1. The relationship between daily mass gain and metabolisable energy intake in suckling llama crias (n=17) at peak lactation (25 days post partum). Predicted maintenance (at zero growth) is 312 kJ kg$^{-0.82}$ d^{-1} (Y = -21.14 + 0.068 X, R^2 = 0.81, P<0.01).

Furthermore, the percentage of water coming from milk decreased as lactation proceeded, while the TWT increased (Table 1), underlining that an increasing amount of water came from other sources as expected. Similar findings were reported for the tammar wallaby (Dove and Cork, 1989) and sheep (Dove, 1988) based on the isotope transfer method (Holleman *et al.*, 1975). When TWT was expressed as percentage of body mass or on the basis of ml per litre body water$^{0.82}$ (Table 1) to discount for differences in body fat content (King, 1979), the same trend as in TWT could be observed.

Dams

As milk production in llama dams decreased with prolonged lactation, water intake and fractional turnover rate also decreased (Table 2), as less water was needed for milk production.

Body water expressed as percentage of body mass (body water fraction) was for all three lactation stages in the range of reported data on adult llamas (Marcilese *et al.*, 1994; Fowler, 1989), indicating that the body water fraction of lactating llamas ranges between 63 to 66% (Table 2). It is interesting to note that the milk water fraction, which describes the relationship between milk-water secreted and TWT, is fairly stable between measurement periods (Table 2), suggesting that during lactation total TWT is directly related to milk production.

Conclusion

The present summary of published data on milk intake and output in llamas using different measurement techniques such as the DTTM and the ID method, gives reasonable estimates for the energy supply for suckling llamas. However, a bigger database is needed for obtaining more reliable estimates on nutrient requirements for lactating and suckling llamas. Until then, the present data on milk and nutrient intakes can serve as a basis for further recommendations.

References

Butte N.F., W.W. Wong, B.W. Patterson, C. Garza and P.D. Klein, 1988. Human-milk intake measured by administration of deuterium oxide to the mother: a comparison with the test-weighing technique. American Journal of Clinical Nutrition, 47: 815-21.

Carmean B.R., K.A. Johnson, D.E. Johnson and L.W. Johnson, 1992. Maintenance energy requirement of llamas. American Journal of Veterinary Research, 53: 1696-1698.

Coward W.A., M.B. Sawyer, R.G. Whitehead, A.M. Prentice and J. Evans, 1979. New method for measuring milk intakes in breast-fed babies. Lancet, 2: 13-14.

Dove H., 1988. Estimation of the intake of milk by lambs, from the turnover of deuterium- or tritium-labelled water. British Journal of Nutrition, 60: 375-87.

Dove H. and S.J. Cork, 1989. Lactation in the tammar wallaby (*Macropus eugenii*). I. Milk consumption and the algebraic description of the lactation curve. Journal of Zoology, 219: 385-397.

Fowler M.E., 1989. Medicine and surgery of South American camelids: Llama, Alpaca, Vicuna, Guanaco. Iowa State University Press, Ames, IA, USA.

Haisma H., W.A. Coward, E. Albernaz, G.H. Visser, J.C.K. Wells, A. Wright and C.G. Victoria, 2003. Breast milk and energy intake in exclusively, predominantly, and partially breast-fed infants. European Journal of Clinical Nutrition, 57: 1633-1642.

Holleman D.F., R.G. White and J.R. Luick, 1975. New isotope methods for estimating milk intake and yield. Journal of Dairy Science, 58: 1814-21.

Holleman D.F., R.G. White and J.R. Luick, 1982. Application of the isotopic water method for measuring total body water, body composition and body water turnover. In: I.A.E. Agency (ed.), Use of tritiated water in studies of production and adaptation in ruminants. International Atomic Energy Agency, Nairobi, Kenya, pp. 9-32.

Johnson L.W., 1994. Lama nutrition. In: L.W. Johnson (ed), The Veterinary Clinics of North America. Food Animal Practice Vol. 10, W.B. Saunders Co., Philadelphia, PA, USA, pp. 187-200.

King J.M., 1979. Game domestication for animal production in Kenya - Field studies of the body-water turnover of game and livestock. Journal of Agricultural Science, 93: 71-79.

Marcilese N.A., M.D. Ghezzi, M.A. Aba, R.A. Alzola, H. Solana and R.M. Valsecchi, 1994. Physiological studies in the South American camelid llama (*Lama guanicoe f. d. glama*). I. Body water spaces and water turnover. Acta Physiologica, Pharmacologica et Therapeutica Latinoamericana, 44: 36-42.

Oftedal O.T., 1981. Milk, protein and energy intakes of suckling mammalian young: a comparative study. Cornell University, Ithaca, NY, USA.

Pouillon C., 2001. Untersuchungen zum Mutter-Kind-Verhalten von Lamas (*Lama glama*) unterbesonderer Berücksichtigung des Saug- und Säugeverhaltens. Justus-Liebig-Universität Gießen, Gießen, Germany.

Riek A., 2008. Relationship between milk energy intake and growth rate in suckling mammalian young at peak lactation: an updated meta-analysis. Journal of Zoology, 274: 160-170.

Riek A. and M. Gerken, 2009. Milk intake studies in llamas (*Lama glama*) using the 'dose-to-the-mother' technique. Small Ruminant Research, 82: 105-111.

Riek A., M. Gerken and E. Moors, 2007. Measurement of milk intake in suckling llamas (*Lama glama*) using deuterium oxide dilution. Journal of Dairy Science, 90: 867-75.

SAS User's Guide: Statistics, Release 9.01. 2001. SAS Inst. Inc., Cary, NC, USA.

Schneider W., R. Hauffe and W. van Engelhardt, 1974. Energie- und Stickstoffumsatz beim Lama. In: Y. van der Honing (ed.), Energy Metabolism of Farm Animals. EAAP publications, Stuttgart, Germany, pp. 121-130.

Van Saun R.J., 2006. Nutrient requirements of South American camelids: A factorial approach. Small Ruminant Research, 61: 165-186.

Feed preferences and recipe alternatives for alpacas in a Hungarian zoo

A. Prágai[1], A.K. Molnár[2], J. Pekli[1], R. Veprik[3], GY. Huszár[1] and Á. Bodnár[1]
[1]International Development and Tropical Department, Szent István University, Faculty of Agricultural and Environmental Sciences. Páter Károly u. 1, 2100 Gödöllő, Hungary; bodnar.akos@mkk.szie.hu
[2]Research Institute for Animal Breeding and Nutrition, Gödöllő Research Institute, Isaszegi u. 200, 2100 Gödöllő, Hungary
[3]Szeged Zoo, Cserepes sor 47, 6725 Szeged, Hungary

Abstract

A feed preference trial was done with four alpacas (1 male and 3 female) in Szeged Zoo, Hungary. Alpacas in Hungarian zoos have adapted well to the local climate, and also show very good results in reproduction. Feeding of these animals is based on local forages, but unexpected feeding problems appear in many cases according to the energy and protein needs. Alpacas are usually fed by wet feeds and forages (apple and other fruits, carrot, cabbage, available fresh feeds) and additionally horse fodder in the local zoos. But it does not completely satisfy the nutrient and protein needs of these animals. Results show that the feeding of alpacas can be based on local forages. Alpacas are thought to particularly like the fresh or wilted feed, and the different kinds of hays were also appreciated by the animals. Based on the results of preferences trials and nutritional parameters, alfalfa hay and grass hay are the basic forage throughout the year. Additionally, in each season there are different additional feeds (e.g. Sudan grass and other fresh forages in spring; apple and cabbage in summer; sugar beet and red beet in autumn; apple, carrot and cabbage in winter, etc.). By harmonising the nutritional parameters of the different forages and the nutritional needs of alpacas, a specific alpaca recipe for zoos can be made.

Keywords: alpaca, zoo, feeding, feed preferences

Introduction

Alpacas are kept only in zoos and wild animal parks as exotic animals in Hungary. However, due to the production of wool for high quality textiles, alpacas can also be kept as livestock animals. The feeding system of alpacas kept in Hungarian zoos is actually *ad hoc*, and based on the essential and relatively low-cost feedstuffs which are available in the actual season. That is the main reason why the feeding of alpacas needs to be considered according to the actual economical circumstances, season and the physiological status of the animals.

The aim of this project was to investigate the feed preferences of alpacas and to find different kinds of feedstuffs to complement their present fodder. We are planning to work out an optimal recipe for alpacas in Szeged Zoo. It is essential to ensure nutritious and sufficient feed for the alpacas, to maintain their health, welfare, reproduction rate and wool production. One of the key financial challenges for a breeding farm or a zoo is to obtain fodder for the animals that meets the qualitative and quantitative requirements. Based on the results of this research, an adequate recipe has been worked out for alpacas, considering the nutritional needs, feed preferences as well as the economic aspects.

Material and methods

Animals and housing circumstances

Four alpacas (3 mares and 1 stallion) were used for this trial at Szeged Zoo, Hungary. Johnny (♂) is four years old and his body weight is 74 kg, while Juanita (♀) is 7 years old and 70 kg, Lilla (♀) is 5 years old and 69 kg, and Szöszke (♀) is a 4-year-old mare with 67 kg. The stable of the alpacas is an 80 m² straw-bedded wooden house with direct connection to the open yard. The yard is approx. 2,500 m² covered by natural pasture and trees, with a small lake in the middle. The botanical composition of the yard is the following: *Lotus tenuis* (dominant), *Stenachtis annua* (10-15%), *Rumex acetosella*, *Achillea distans*, *Trifolium pratense*, *Papaver rhoeas*, *Plantago lanceolata*, *Calystegia sepium* and *Urtica dioica*. The dominant grass species is *Cynodon dactylon*, but *Lolium perenne*, *Arrhenatherum elatius*, *Bromus inermis*, *Phleum pratense* and *Phragmites australis* can also be found in high percentages. The nutrient content of the herbaceous plants in the yard changed during the vegetation period, as shown in Table 1.

In addition, there were two plants of *Olea europea* and several plants of *Tilia cordata* and *Cynosbati pseudofructus* in the yard. Alpacas readily consumed the leaves of these arboreal plants to supplement their daily fodder.

Procedure of the trial

Portions of the different feedstuffs (4 similar kind of feeds, 1 kg from all at the same time) were laid at the entrance of the yard, next to the gate where the animals left their stable. Freedom of choice for the alpacas was ensured by creating a 1-1.5 m distance between the different portions.

According to Alonso-Díaz (2008), the total duration of the investigation was 4 hours per day. As the literature shows, feed preference investigations were repeated on three days following each feedstuff (Scott, 1995). Typical and available feeds in the different seasons in Hungary were chosen for this trial. Feed preference trials were done according to the selected feedstuffs listed in Table 2. Dry matter, protein, fat, fibre and ash content of the listed fodders have been defined. The amount of time the alpacas spent on consuming the different fodders was examined each day of the experiment. Furthermore, the number of occasions on which animals fed on the single fodders was also observed. Two observers were responsible for recording the data during each investigation day.

Table 1. Changing of the grass forage ingredients of the yard during the vegetation period.

	Beginning of the vegetation period (May)	End of the vegetation period (September)
Dry matter (g/kg)	222.2	444.5
Crude protein (g/kg dry matter)	200.3	155.7
Ether extract (g/kg dry matter)	53.1	36.4
Crude fibre (g/kg dry matter)	240.1	334.9
Ash (g/kg dry matter)	97.6	120.1
Nitrogen-free extract (g/kg dry matter)	414.8	452.9

Table 2. Different feedstuffs used for feed preference trials.

Season	Feeds	Date
Winter	Alfalfa pellet, grass hay, alfalfa hay, corn silage	20-22/02/2009
	Apple, carrot, cabbage, fodder beet	25-27/03/2009
	Wheat, barley, corn, oat	06-08/03/2009
Spring	Alfalfa, grass, pea stem, lettuce	24-26/04/2009
	Barley/sunflower (meal), soybean meal, ground rape, barley/ sunflower (pellet)	01-03/05/2009
Summer	Vetch and oat, sunflower, corn, sorghum (all of them are parched)	19-21/06/2009
	Courgette, melon, marrow, sorrel	17-19/07/2009
	Leaf of *Robinia pseudoacacia, Celtis australis, Tilia cordata* and *Elaeagnus commutate*	07-09/08/2009
Autumn	Pumpkin, sugar beet, beetroot, potato (all of them are chipped)	11-13/09/2009
	Calf nutriment, milking nutriment, flake corn, sheep nutriment	24-26/09/2009

Results and discussion

Feed preference investigations

Trials in winter

The alfalfa pellet seemed to be most appreciated by the three mares and they ate all of this in a very short time. The corn silage was not eaten by the animals very much. The mares smelled it many times, but only the stallion consumed it. Alfalfa hay was liked by the male, while the mares preferred the grass hay. The three mares did not seem to be choosey regarding the apple and vegetables. They took from all of the vegetables from the very first day of the trial. While the stallion only consumed the fodder beet. The animals did not eat fodder during the 3 days of the examination (wheat, barley, corn and oat). The duration of consumption from different feedstuffs is shown in Figure 1.

Trials in spring

Feed preference results from spring are presented in Figure 2. As the figure shows, the mares liked the dried fodders more than the stallion. There were major differences in the choice of roughage: Lilla consumed alfalfa and pea stem (with higher protein content) much more than the others, while Juanita preferred the more succulent feeds with smaller protein content (grass, lettuce). Meals were only smelled by the animals, but the alpacas did not consume these fodders at all.

Trials in summer

Feed preference results from summer are presented in Figure 3. The alpacas liked the leafy branches (cut off from the trees) very much. Oat vetch, corn, sorghum and sorrel were next in the preference list. They liked sunflower mixed pickles less because of their hairy surface, while the courgette and pumpkin was not really favoured by the alpacas due to their unfamiliar flavour.

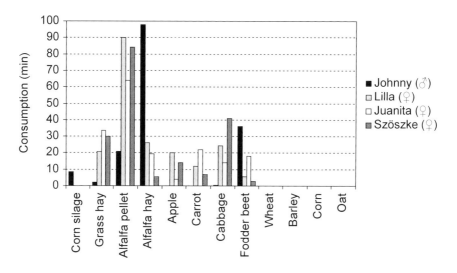

Figure 1. Consumption time of different feedstuffs in winter.

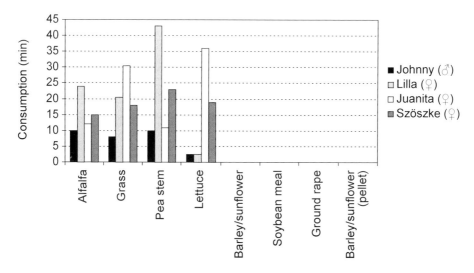

Figure 2. Consumption time of different feedstuffs in spring.

Trials in autumn

Feed preference results from autumn are presented in Figure 4. As the figure shows, alpacas were not curious about the ready fodders (calf, milking, sheep nutriment) and the flake corn. This is a potentially interesting observation, because the alfalfa pellet with a similar appearance was a much preferred fodder in winter. It was found that alpacas greatly liked the root tubercle fodders.

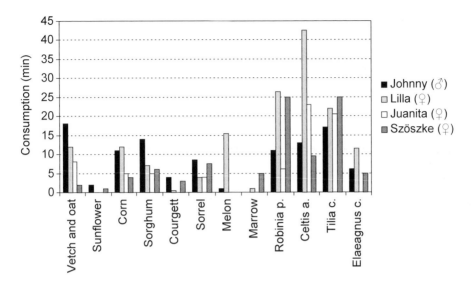

Figure 3. Consumption time of different feedstuffs in summer.

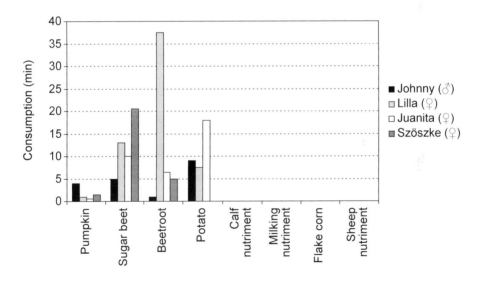

Figure 4. Consumption time of different feedstuffs in autumn.

Feeding programme

Following the feed preference examination, we drew up the feeding proposal with the preferred feedstuffs for average weight of 70 kg alpacas. It was done by considering the following:
1. Van Saun (2005) proposed the daily nutrient requirements of the full-grown alpaca (with a weight of 70 kg): dry matter 1.26-1.4 kg (1.8-2% of the body weight), protein 150-170 g (120 g/kg of dry matter) and fibre 25-30%.

2. There is no pasture in winter time, therefore hay and succulent fodders (root tubercles, fruits) could be given in this season.
3. Dry matter content of the pasture grass is low in spring. It is necessary to supplement with hay at this time.
4. Because of the drought and the ageing of the pasture grass, supplementation with alfalfa hay, leafy branches (protein source) and succulent fodders is proposed at the end of summer.
5. It is recommended to feed fruits and root tubercles (carrot, beetroot) to ensure good vitamin levels in the animals in autumn.
6. The basic fodder has to be available the whole year round at a favourable price and in large volumes. To provide enough 'green forage', grazing or cultivated green plants are recommended which can be obtained in large quantities (sorghum, oat and vetch, corn mixed pickles). When there is no 'green forage', it is recommended to feed the animals with grass and alfalfa hay, apple, carrot, potato, beetroot, cabbage.
7. It is strongly recommended to consider the chemical composition of the preferred fodders.
8. It is necessary to provide fresh, clear, drinking water supply at a suitable temperature all year round.

The alpaca feeding programmes which can be applied at a reasonable cost in Hungary in winter, summer and in the transitional periods between them are summarised in the following tables (Tables 3-5).

Table 3. Minimum feed needs for a 70 kg alpaca in winter.

	Quantity of feedstuff (kg)	Dry matter intake (kg)	Protein intake (g)	Fibre intake (g)
Grass hay	1.7	1.5	120	540
Alfalfa hay	0.3	0.26	40	91
Apple+carrot+cabbage+beetroot	0.5[a]	0.05	5	3
Potato	0.3	0.06	5	2
Sum	2.8	1.87	170	636 (34%)

[a] This quantity contains all of the quantities of the available feedstuffs.

Table 4. Minimum feed needs for a 70 kg alpaca in summer.

	Quantity of feedstuff (kg)	Dry matter intake (kg)	Protein intake (g)	Fibre intake (g)
Grass[1]/'green forage'[2]	3.6	1.30	210-160[1]	455
Grass hay	0.4	0.35	28	122
Sum	4	1.65	238-188	577 (35%)

[1] From Table 1.
[2] Grass, 'green forage' mixture (spring: vetch and oat, grass and pulses; autumn: vetch and wheat, etc.) and sorghum (high quantity in a short time, palatable, high nutrients content).

Table 5. Minimum feed needs for a 70 kg alpaca in the transitional periods (6 weeks).

Feedstuff (kg)	Winter	Transition periods						Summer
		Week 1	Week 2	Week 3	Week 4	Week 5	Week 6	
Grass hay	1.7	1.7	1.5	1.3	1.0	0.8	0.6	0.4
Alfalfa hay	0.3	0.2	0.2	0.1	0.1	-	-	--
Fruits	0.5	0.4	0.3	0.2	0.1	-	-	-
Potato	0.3	0.3	0.3	0.2	0.2	0,1	-	-
Grass/'green forage'	-	0.5	1	1.5	2	2.5	3	3.6 (70%) [a]

[a] Percentage of the daily dry matter intake.

Conclusions

Based on the results, it is obvious that the tubercles, fruits and vegetables can be given to the alpacas at any time, because the animals can utilise their vitamin and carbohydrate content very well. It is necessary to pay attention to the low dry matter and fibre content of these kinds of fodders when consumed. It is advised to feed these fodders together with fodder with a high dry matter and fibre content (hays, chopped dry corn-stalk, etc.) only.

It was observed that the alpacas preferred the different hays and the alfalfa pellet, but the pellet is a very expensive fodder. Therefore, the grass or alfalfa hay should be the basic fodder because these can provide the animals with the right amount of dry matter, fibre content and crude protein.

The succulent fodders and the fresh green leafy branches can be a good supplement to the basic fodder, mainly in the period when the quality of pasture is very low. This is very important because the mares are pregnant or already in the lactation period at that moment. This forage proposal is calculated for adult, unproductive alpacas. Therefore, in addition to the basic fodder an increased amount of fodder is needed for productive animals (e.g. growing foals, mares in the last third of the pregnancy, lactating animals) (Van Saun, 2006).

References

Alonso-Díaz M.A., J.F.J. Torres-Acosta, C.A. Sandoval-Castroa, H. Hoste, A.J. Aguilar-Caballero and C.M. Capetillo-Leal, 2008. Is goats' preference of forage trees affected by their tannin or fibre content when offered in cafeteria experiments? Animal Feed Science and Technology, 141(1-2): 36-48.

Scott B.C., 1995. Dietary habits and social interactions affect choice of feeding location by sheep. Department of Rangeland Resources, Applied Animal Behaviour Science, 45(3-4): 225-237.

Van Saun R.J., 2006. Nutrient requirements of South American camelids: A factorial approach. Small Ruminant Research, 61(2-3): 165-186.

Management

A cross-border wool project supports the conservation of the Alpines Steinschaf

C. Mendel[1], A. Feldmann[2] and N. Ketterle[3]
[1]Bavarian State Research Center for Agriculture, Institute of Animal Breeding, Prof.-Dürrwächter-Platz 1, 85586 Poing-Grub, Germany
[2]Society for the Conservation of old an Endangered Livestock Breeds in Germany (GEH), Postbox 1218, 37202 Witzenhausen, Germany; feldmann@g-e-h.de
[3]Ark-Farm Ketterle, Bossler Str. 1, 73119 Zell u.A., Germany

Abstract

The Alpines Steinschaf is an old and traditional sheep breed that has lived in the Alps for several hundred years. In 1985 only a few small flocks of this breed could be identified in special alpine regions in Germany. A breeding programme was established by some interested breeders and the official breeding organisations. To maintain rare breeds for the future, it is necessary to develop special programmes for their use and marketing. In 2004 the breeders in Germany and Austria created a programme to market high-quality products made from the wool of this sheep breed. The project has been running very successfully and, since its inception, the population of the Alpine Steinschaf sheep has been increasing annually.

Keywords: conservation, sheep, traditional breeds

Current situation

The Alpines Steinschaf is a direct descendant of the so-called Torfschaf and thus it belongs to one of the original breeds of the European Alps. In 1863 about 208,000 animals existed (Mason, 1967). In some regions, the Alpines Steinschaf was the most popular breed. Animals could be found in the area of Bavaria in the region of Berchtesgaden, Traunstein and Rosenheim (Kaspar, 1928). In Austria this breed was kept mainly in the area of Salzburg and Tirol (Führer, 1911). Exactly 100 years later in 1964 the population had fallen to a meagre 1000 animals (Mason, 1967). In 1985 some interested breeders from Germany and Austria started to collect the last individuals of the breed in the Berchtesgarden area (Mendel and Burkl, 2009). With the help of the Sheepbreeders' Associations in Bavaria and Tyrol a new workgroup was established for the conservation of this specific breed. Between 1990 and 2004 the population increased from 50 breeding animals up to 295 animals (Feldmann et al., 2005). In the study on genetic distances of alpine sheep breeds of the eastern alps it was clear that the Alpines Steinschaf is genetically separated from the other types of Steinschaf like Montafoner-, Krainer- and Tyrol Steinschaf (Baumung and Sölkner, 2003).

The organisations in Germany and Austria

In the area of *in situ* conservation of rare breeds, it is mainly non-governmental organisations (NGOs) that are active. In Germany and Austria these NGOs have been around for nearly 30 years. The main objective in the work with the rare breeds is to co-ordinate the breeders and to inform the public about the loss of genetic diversity in modern agricultural systems.

Without financial support from the state for the breeders of old and traditional livestock breeds, more breeds will disappear in the next few years. The FAO maintains its position that nearly every two weeks, one of the 7,800 breeds which are registered today will be lost. There is no doubt amongst livestock keepers and researchers that for the future adaptation a diverse gene-pool of agricultural livestock will be the main foundation of sustainability and the fight against hunger worldwide.

In Germany the Society for the Conservation of Old and Endangered Livestock Breeds (GEH) was founded in 1981 by a small group of people, who recognised, that some breeds had disappeared from animal fairs, from breed competitions and farmers' stables. A lot of work was done to stop the extinction of breeds. Since the existence of the GEH not a single breed has vanished. The so-called 'Red List of endangered Livestock Breeds' in Germany consist of about 103 breeds from 14 species. With the help of this list it is possible to follow the trend and situation of each endangered breed and to start useful activities to maintain the population. Today the GEH has more than 2,200 members. Since 1984 an 'Endangered Livestock Breed of the Year' has been awarded annually. In 2009 the Alpines Steinschaf was given the title, being one of the most endangered breeds in Germany. In 2010 the Meißner Widder Rabbit was the focus of the activities.

The situation in Austria is quite similar to Germany. The Arche Austria Society has existed since 1982 and works mainly with the breeders to install further flocks of the rare breeds for *in situ* conservation. There is a close collaboration with breeding organisations and research institutes. The Arche Austria also named the Alpines Steinschaf 'Breed of the Year 2009' (Arche Austria, 2009).

Breed description of Alpines Steinschaf

Breeding history

Like all other Steinschaf breeds, the Alpines Steinschaf can be traced back to Zaupel and Neolithic Torf. In former times the breed was widespread in the Alpine region. However, since 1960, it has to a large part been replaced by other breeds and is extremely endangered today.

Description of the breed

The Alpines Steinschaf is a small- to medium-sized finely boned mountain breed. The body is broad and compact. Animals have a noble head with straight profile and a slightly roman nose. The ears are lop and bent slightly forward. Both sexes are generally horned. The rams usually have spirally curved horns, the ewes sometimes small and slightly curved horns (Figure 1). The legs are thin but strong with hard hooves. Their long woollen tails reach down as far as the ankle joint and the ends are often bent. The face, stomach and feet are usually wool free. The forehead is covered with short wool. All colours and markings occur in the rough mixed wool (Figure 2). They mature early and are year-round breeders. Mason (1967) describes the breed as being very fertile with around 20-70% of births involving twins.

Special performances

This breed is robust, long-living and frugal and still able to thrive under extreme living conditions (Table 1). It has adapted to marginal sites and high precipitation, has good maternal instincts and a reduced susceptibility to diseases (foot rot). The quality of the meat is good.

Figure 1. A ram of the breed Alpines Steinschaf. Photo: Feldmann (GEH, 2009).

Figure 2. A group of mothers with their Alpines Steinschaf lambs in different colours. Photo: Milerski (GEH, 2009).

Table 1. Performance of Alpines Steinschaf (Feldmann et al., 2005).

	Body weight (kg)	Fleece weight (kg)	Lambing percentage (%)	Withers height (cm)
Rams	60-75	3.5		73-80
1 year old rams	40-60	3.0		
Ewes	45-60	3.0	180-220	65-70
1 year old ewes	35-45	2.0		

Management

Population size

In Germany 304 ewes and 14 rams are registered in the herdbook. In Austria 189 ewes and 33 rams are registered in the programme of the Austrian Association for Rare Endangered Breeds (ÖNGENE).

Breeding focus is put on the breed's genetic conservation.

Characteristics

Besides meat and wool production, the breed is mainly used for landscape management, especially marginal sites. In Bavaria a programme has been set up to focus on the adaptation of the Alpines Steinschaf to the harsh conditions in the high mountains. For this purpose a group of sheep has been put out to alpine pasture and different research is being conducted to acquire more details about the health and resistance of this breed.

Special characteristics of the Alpines Steinschaf are:
• adaptability to the harsh mountain pastures at heights which are inaccessible to cattle;
• very fertile with ability to breed out of season;
• excellent potential for milk production and dedicated mothers as well as longevity;
• hardy, rugged, resistant and easily satisfied;
• exceptionally tame;
• multifunctional usability, coloured, dual-coated fleece.

Related breeds

The Alpines Steinschaf is related to other breeds in the alps like Krainer Steinschaf in Germany and Austria, Bovsca Ovca in Slovenia, Plezzana in Italy and Montafoner and Tyrol Steinschaf in Austria.

Government support for Alpines Steinschaf

The aim of government support for rare breeds is to establish a breeding programme which helps to increase the number of animals, to prevent inbreeding and to give the breeders economic support.

The support of rare breeds is not only given on a national level, but also on a European level. The European Union supports animal genetic resources based on EC regulation 1257/99 for the protection of rural development. Under this regulation, EU countries support rare breeds in the framework of their plans for the development of rural areas.

In Germany funds for traditional breeds are provided by the ministries for agriculture of the federal states. For breeders of the Alpines Steinschaf funds are given only in Bavaria, the state of the origin of the breed. This allocation is based on the directive of the Bavarian Ministry (AllMBl. Nr. 13/2008) for animal genetic resources. The breeder can get €20.00 for each registered animal, with a minimum €100.00 per farm, and a maximum €2,000.00 per farm. Five years of contribution to the programme is required.

In Austria the Association for Rare Endangered Breeds (ÖNGENE) calculates the support for rare breeds. The support is co-financed by the European Union and is part of the development of rural areas. There is the possibility to support all farmers who take part in the Austrian programme for environmentally friendly agriculture and who keep listed endangered breeds. These subsidies do not increase production, but they represent approval for the difficult breeding work. At the moment there are basic subsidies and supplements for highly endangered breeds, which are included in the OPTIMATE programme. For the Alpines Steinschaf €55.00 for ewes and €120.00 for rams are given per year to breeders who register their sheep in the ÖNEGEN breeding programme.

The Alpines Steinschaf wool project

Wool as a niche product

The long-term security of flocks depends on the economic value of a breed. A combination of governmental support for breeding animals and for environmental services helps to maintain rare breeds in agriculture. In addition to this, the farmers themselves have to think about niche production with the animals and their products. This means that quality products have to be produced to place them on the market in the higher price segments. It is not only the yield of meat or the yield of wool that generates profit for the breeders. Products with a special quality often fetch better prices.

The wool of the Alpines Steinschaf is a product that is specific to this breed and therefore can be used to develop a special type of marketing. For years the price for the wool of landraces like Alpines Steinschaf has been very low on the market, not more than €0.10/kg. The cost of shearing one sheep is around €3.00. This means that the breeder has to pay much more for the shearing than he gets for the wool. The aim of the wool project is to give the breeders a fixed price of €1.00/kg for wool of high quality.

Quality of the wool

The beautiful and unique wool of the 'Steinschaf' is a distinguishing feature of this ancient and rare breed. It has a dual-coated fleece with pithy, long coarse hair, fine wavy and short bottom hair, and every wool colour from white, black through to browns and brindled can be found. Ewes produce about 3.0 to 3.5 kg and rams 4 kg (Feldmann et al., 2005).

The wool project

The breeders themselves wanted to be active. In 2004 the German and Austrian breeders joined forces and organised a workgroup to make high quality products out of the wool. As a result of the still small weight of wool being collected by the breeders, the big enterprises involved in processes like washing or combing the wool refused to accept the small amount of wool.

The first attempts in 2005 to sell the products directly to small regional markets or small markets on the farms of the breeders were successful. The presentation of different products like socks, sweaters and knitting wool won over a lot of customers. As a result of this experience, at their annual meeting the breeders discussed the next step and the commercialisation of the products began.

Collecting the wool

More and more breeders were convinced and started to collect wool from their sheep. At the annual meeting in May each year wool is gathered and sorted. As a consequence nearly two tonnes of wool from the Alpines Steinschaf have been collected since its beginning in 2004.

From 2004 to 2009 the number of breeders increased from 2 to 18. The amount of wool collected also grew from 111 kg in the year 2004 to 730 kg in 2009 (Figure 3). Altogether, about 2 tonnes of wool of the Alpines Steinschaf was available for the manufacture of products with high quality characteristics. The cooperation of the breeders therefore affects the rise in wool production.

Breeders receive a wool voucher of €1.00 for each kilogram of prepared wool. The aim is to raise the prices further in order to make it more profitable for the breeders. The wool quality becomes a very important breeding aspect, especially the natural colouring. In addition to this length and density, stability, refinement and crimps of the wool are included in the breeding aims. The economic value of the wool protects the continuity of the breed. Collecting raw wool at the annual meetings means that transport costs can be kept very low (Figure 4). Via the wool project the wool is washed and refined (combed, spun, knitted, woven, sewn, etc.). 100 kg of raw wool yields approximately 60-70 kg of washed pure wool. Depending on the wool type, 30-40% of the weight is wool grease and dirt (ERSA, 2007).

Processing of the wool at regional manufactories

One very important aim of the breeders was to work together with traditional and local craftspeople for this project. This processing is now carried out exclusively in Germany and Austria and only at small firms or manufacturers. Transparency in the production process and the fact that the products are unrefined is of great importance for the breeders.

Finding companies that are prepared to take on small amounts of wool and are able to produce high-quality products is not easy. In the textile sector as well as in the agricultural sector and other related branches there has been a strong focus on specialised firms producing their products in other countries.

A lot of networking was required to find small regional enterprises that could wash and clean the wool. The next step was to find a factory that would produce the products, while the number of socks and sweaters were still in small numbers. Two enterprises were found in Austria to produce the quality the breeders were looking for.

The Alpines Steinschaf current production range is made up of combed wool, knitting wool, woollen socks, gloves and mittens, jumpers, sweaters. Felt products include bags, felt insoles, felt slippers, cushions, etc. Around 40 different products in different colours are designed. The

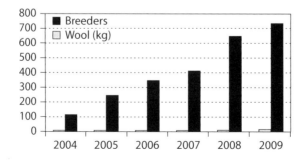

Figure 3. Evolution of breeders and wool (kg) from the start of the project in 2004 up to 2009 (personal communication, 2009).

Figure 4. The wool has to be sorted and put into big bags for transportation. Photo: Ketterle (GEH, 2009)

breeders themselves sell the finished products to their own customers. The range of high quality products is increasing annually with consumer demand for contemporary wool clothing.

Crucial for the Alpine Steinschaf breeder, however, is the regional production of high-quality products without cheap mass-production. The products from the 'Alpine Steinschaf' range enrich the produce range of farm shops, market stalls as well as fairs and exhibitions making the selling of these products interesting for dealers (Figure 5). Fortunately, more and more consumers are aware of high-quality natural products and are prepared to pay the price.

Figure 5. Selling products at the International Green Week in Berlin. Photo: Ketterle (GEH, 2009).

The wool project helps to maintain the Alpines Steinschaf

Conversely, the products promote the Alpines Steinschaf. This is especially important as only with increased awareness, popularity and better marketing of the animals and its products is it possible to bring this breed back from the brink of extinction and thus preserve an object of cultural value.

A very attractive Internet platform at http://www.alpines-steinschaf.de shows the activities and products of the Alpines Steinschaf. The breeders together decided to create a labelled product (Figure 6).

Figure 6. Logo on wool products from the Alpines Steinschaf.

Conclusion

The Alpines Steinschaf is one of the most endangered sheep breeds of Germany and Austria. Since 1985 various activities have been initiated to preserve this alpine breed. In 2004 some breeders began to collect the typical wool of the breed to produce high-quality products from the wool and to guarantee fixed prices for the breeders. They started to go to regional markets and sell the products at good prices. The activities were conducted around the following objectives:
- buying of pre-sorted wool at a fair price;
- organic processing and production of high-quality wool products;
- regional production which supports local trade;
- processing under socially acceptable working conditions;
- supporting agriculture and landscape conservation through sheep holdings;
- promotion of an old and endangered sheep breed;
- conservation of genetic diversity and an object of cultural value.

References

ARCHE Austria, 2009. Rote Liste der gefährdeten Haustierrassen in Österreich. Available at: http://www.arche-austria.at.
Baumung R. and J. Sölkner, 2003. Genetische Differenzierung von Schafrassen im Ostalpenraum. Final report, Division of Livestock Sciences, University of Natural Resources, Vienna, Austria, S. 39.
ERSA, 2007. Alpinet Gheep Alpine Network for sheep and goat promotion for a sustainable territory development. Agenzia Regionale per lo Sviluppo Rurale Friuli Venezia Giulia (ERSA), pp. 100-110.
Feldmann A., U. Bietzker and C. Mendel, 2005, Schafrassen in den Alpen, Gesellschaft zur Erhaltung alter und gefährdeter Haustierrassen e.V. (GEH). Bayerische Landesanstalt für Landwirtschaft (LfL), pp. 8-9.
Führer L., 1911. Studien zur Monographie der Steinschafe. University of Natural Resources, Vienna, Austria.
GEH, 2009. Rote Liste der gefährdeten Nutztierrassen in Deutschland. Available at: http://www.g-e-h.de.
Kaspar K., 1928. Studien über das Steinschaf im Chiemgau. Thesis, Technical University Munchen, Munchen, Germany.
Mason I.L., 1967. The Sheep Breeds of the Mediterranean. Commonwealth Agriculture Bureaux, Edingburgh, Scotland, UK.
Mendel C. and G. Burkl, 2009. Gefährdete Nutztierrassen Schwerpunkt Schafe, Ziegen, Gebrauchshunde, Gesellschaft zur Erhaltung alter und gefährdeter Haustierrassen e.V. (GEH), pp. 5-6.

Investing in the development of South American campesino camelid economies: the experience of the International Fund for Agricultural Development (IFAD)

R. Haudry De Soucy
International Fund for Agricultural Development (IFAD), Via paolo di Dono 44, 00142 Rome
Italy; r.haudry@ifad.org

Abstract

The Andean region has three million domesticated South American camelids and one million in the wild, a valuable asset with underused potential for contributing to climate change adaptation, high-quality fine-fibre textiles, and the meat, leather and other industries. Camelid raising is unique in two respects: it is practised mainly at altitudes of 3,000 meters and above, in the fragile ecosystems where it evolved and to which it is perfectly adapted; and it is an asset of the poorest people living in this region. In short, this is essentially an activity of campesinos in the Andean highlands. In the past 20 years, the International Fund for Agricultural Development (IFAD) has invested in bringing together knowers and knowledge. Breeders, fibre sorters, artisans and microentrepreneurs linked to the camelid economy have exchanged and scaled up knowledge, with tangible results. IFAD has also cofinanced their enterprises and initiatives with support from national and local governments, businesses and other sources of cooperation. An estimated US$30 million has been invested in the sector in the past 15 years, generating an impact or incremental value of US$90 million in the pockets of producers. These are only the first steps in the process of developing the camelid economy and its contribution to mitigating climate change and combating rural poverty. In the four countries concerned, the sector could generate three times the current annual sales (US$300 million in the campesino sector and US$1,000 million if one includes the textile and garment industries and livestock sales). Such an increase will hinge on expanding incipient markets for meat (dried and fresh), leather manufactures and livestock by-products, as well as the textile market for vicuña fibre (guanaco fibre is still virtually unused). Knowledge exchanges need to be multiplied among local and international talents, because the thresholds for learning simple technologies and prerequisites to access solvent markets are beyond reach for many campesinos and artisans. This calls for cofinancing knowledge exchanges between subregions and countries, scaling up such knowledge to the majority, and selectively cofinancing the most promising and environmentally relevant initiatives to contribute positively to climate change (for instance, by substituting camelids for exotic livestock such as cattle and sheep in moist paramos or high steppe areas).

Keywords: development, camelid economy, campesino

Introduction

With six million South American domestic camelids and one million wild ones, the Andean region possesses a substantial quantity of livestock that is underutilised in terms of its contribution to climate change mitigation, the supply of high-protein foodstuffs, food security and the production of inputs indispensable to the world trade in fine textile fibres, meat, leather, etc. These animals also have two unique features:

1. They are mainly reared at altitudes of over 3,000 metres in fragile ecosystems without damaging these, inasmuch as they have evolved there and are thus perfectly adapted to the environment.

2. They are assets belonging almost exclusively to the communities with scant resources who inhabit this region; in other words, camelids constitute a type of livestock and a key economic sector for the sustainable use of high Andean campesinos' resources.

Main features and rationale of the camelid economy: campesinos, ranchers, craft workers and small rural entrepreneurs

It is estimated that some 120,000 high Andean families obtain part of their home consumption and monetary income from the sale of wool, meat and other by-products from this livestock. More than half of these families are classified as extremely poor and the remainder have incomes of less than US$2 per day, while a further 25,000 families, most of them also poor, live from the primary processing of meat, wool and skins, and allied activities.

In their subsistence strategies, a total of approximately 600,000 people depend on this economic sector, which is marked by very low quantities of public and private investment.

They have varying numbers of animals and varying economic and sociocultural conditions. Most of the 120,000 producer families (some 90,000) have fewer than 20 camelids, complementing these with sheep, goats and occasionally cattle, and with highland farming (basically quinoa and bitter yam). Camelids are their source of meat for home consumption, wool for sale in an unprocessed state and yarn to make their own clothes, while they also use dung as fertiliser for quinoa and as cooking fuel. In some zones, llamas are still occasionally used to transport small loads. These animals can represent up to 25% of the communities' income (combining home consumption and sales). Most of their monetary income, however, comes from the sale of their labour outside the family holding, sometimes thousands of kilometres away.

A further 30,000 families own more animals (a simple average of 100) and are those who can be termed livestock breeders or ranchers, inasmuch as they invest in the care of their livestock (especially alpacas, whose main economic purpose is being sheared for wool). Some of these families may have as many as 1000 head of livestock, but they are the exception. Many of them are organised into communities, associations, cooperatives and enterprises. Most of these producers also pursue other economic activities focusing on agriculture, livestock rearing, petty trade and craft work, while a few have launched into tourism-linked activities. Their average annual income from the sale of wool may be up to 40 per cent of their monetary income. Most of them are considered statistically as poor[3].

The remaining 25,000 families that do not directly own livestock (or that tend it for third parties) pursue processing activities based on camelid by-products, especially the production of fabric from homespun or machine-spun yarn[4] and the trade in meat, leather and other by-products. Half of them live below the poverty line. However, these craft workers and small entrepreneurs are in the forefront of value addition processes regarding the sector's products, and are also those who build up relationships with the textile industry and restaurants, act as the driving force behind new market segments such as associative enterprises dealing in dried meat, small factories producing salami-type products, manufacturers of up-market leather goods and selective meat markets (for example hospitals, because of the intrinsic qualities of this meat, such as its low

---

[3] An average producer with some 50 head of alpacas receives approximately US$400 per year for wool – or US$1 per day per family.

[4] Over the past 15 years, a growing number of associated wool producers have been sending the raw material to spinning mills to increase its added value, then selling the spun yarn or making clothes in family enterprises.

cholesterol content), and seek to improve the hygienic quality of their products in order to place them in markets with greater added value.

The distribution of these people among the various countries is similar to that of the livestock itself, so that the vast majority of llama owners are found in Bolivia and the largest numbers of alpaca owners in Peru. In economic terms, the distribution of the primary production value of the camelid economic sector is estimated at 60% in Peru, 30% in Bolivia and the remainder in Argentina, Chile and to a very small extent Ecuador (less than 1%), while the invoicing of the high-end textile industry is concentrated in Peru (75%).[5]

Among the most promising market segments, the management, controlled shearing and sale of vicuña wool should be particularly noted. This relatively new activity represents an estimated annual income of US$10 million for poor campesinos and members of village communities (around 20 thousand families), and is steadily expanding thanks to international regulations and agreements, which have led to good management and a significant increase in numbers of this wild species. While this development has positive environmental effects, it also leads to conflict with traditional livestock farmers and poachers.

In the past twenty years, the International Fund for Agricultural Development (IFAD) has invested resources in collecting the knowledge of the most successful families and entrepreneurs in order to share and scale up knowledge among experts, professionals, breeders/producers, craft workers and microentrepreneurs involved in the camelid sector. This has been made possible through internships and exchanges, and also through schemes to cofinance various endeavours and initiatives, with the support of national and local governments, enterprises and other sources of cooperation.

The results have been tangible. For example, thousands of families are now selling graded rather than ungraded wool, shearing is being mechanised, animals are being slaughtered in abattoirs, the prices of fresh meat and the hygienic conditions of its place of sale are improving, dried and salted meat is being processed in small factories run mainly by women and is being sold in supermarkets in such cities as La Paz, dishes based on camelid meat are being offered in most first-class restaurants (llama meat in Bolivia and alpaca meat in Peru), many livestock farmers' organisations are now selling part of their wool in spools of industrial yarn in an adequate range of colours and are thus obtaining substantial increases in income, and campesinos are obtaining cash from the sale of wool and vicuña and are seeking to process it in order to obtain added value. These changes seemed impossible only a few decades ago, but are steadily expanding today, leading to constant innovations that require ever-increasing investments.

As one of the sources that has contributed to these changes, IFAD[6] estimates that it has directly invested some US$30 million in the sector, with an impact or incremental value of some US$90 million as cash in producers' pockets (in recent years the aggregate incremental income from these investments for the Andean region is estimated at some US$6 million per year).

In terms of the development of this economic sector and bearing in mind its contribution to climate change mitigation and rural poverty eradication, these are in fact only the first steps in the process. In the five countries concerned, the sector could annually invoice three times more than

---

[5] Depending on the year and on wool prices, in Peru this ranges from US$75 million to US$200 million per year. The main importers of wool in tops and spun yarn are China and Italy.

[6] IFAD is a financial institution of the United Nations specialising in the eradication of rural poverty (see the website www.ifad.org).

is at present the case (some US$140 million in the campesino sector and some US$700 million if the amount invoiced by the textile and fashion industry, restaurants and livestock dealers is included, Table 1).

What incentives and conditions does the camelid sector economy require in order to optimise the value of its activities?

With a view to boosting the sector, the embryonic markets for meat (both fresh and dried), llama wool (still little used in comparison with alpaca wool), leather goods and livestock by-products need to be expanded, as does the market for such new textile products as vicuña and guanaco wool (the second of which is still barely being optimised). The exchange of knowledge among local and international local talents must be stepped up, inasmuch as few campesinos and craft workers have yet reached their learning 'ceilings' regarding simple technologies and the minimum qualities needed to enter profitable markets.[7]

This requires that: (1) the exchange of knowledge among subregions and countries be cofinanced; (2) this knowledge be scaled up so that it becomes generalised; and (3) the most promising and environmentally relevant initiatives be selectively cofinanced in order to help counteract climate change (for example, by replacing exotic livestock – cattle and sheep – with camelids in moist highland zones).

The most successful experience is scaled up on the basis of a mapping of good practices and local talents, internships (residential courses and face-to-face exchanges for three or four days among producers and in markets), learning routes,[8] fairs and meetings. The creation of a favourable environment requires the application and enforcement of current regulations governing such

Table 1. Camelids and estimates of their economic values in the Andes for campesinos and microentrepreneurs (millions of US dollars per year).

Species	Head of livestock	Direct campesino income	Textile added value[1] (small enterprises)	Other added value (skins, leather, etc.)	Total economic value received by families in the sector
Llamas	3,000 000	33	3	6	42
Alpacas	3,000,000	45	30	15	90
Vicuñas[2]	350,000	3	7	0	10
Guanacos	750,000	0	0	0	0
Total	7,100,000				142

[1] Homespun yarn, selection and grading of wool, and small processing enterprises owned by campesinos or craft workers in the sector. Does not include the textile industry.
[2] There were only 160,000 vicuñas in 1999.

[7] The gap between the average producer's technological and commercial level and that of best practices is huge and could be narrowed through a sustained exchange of knowledge and access to information and markets, combined with very small cash investments (less than US$500 per family on average).
[8] See the PROCASUR website www.procasur.org.

subjects as poaching, the slaughter of animals, cuts of meat, and public investment in livestock health and in the elimination of obstacles to domestic and international trade.[9]

Similarly, extensive campaigns are needed regarding the benefits of camelids for natural grazing lands and the environment, and also the benefits of their by-products, particularly meat. In this connection, states, enterprises and local governments should allocate greater resources with the twofold aim of generating income and enhancing resources in the structurally poorest zones of the Andean region, while increasing the supply of new products and services to society as a whole.

Through its programmes, IFAD has identified local talents who quickly pass on their knowledge, and has also cofinanced hundreds of small-scale initiatives using a competition-based methodology for the allocation of funds. This has been seen particularly in Bolivia and Peru, where there are a number of investment projects under way in support of the initiatives of campesinos, microentrepreneurs and their governments.

---

[9] A particularly serious case of bad trade practices severely impedes the activities of wool producers in Bolivia who seek to use the facilities of the Peruvian textile industry and have to meet complex bureaucratic requirements to send their wool to Peru and then receive it back in the form of yarn, paying duties and bribing border authorities in order to carry out these value-adding processes.

Building and scaling up knowledge on camelids

G. Vila Melo[1] and C. Gutiérrez Vásquez[2]
[1]Fundación Biodiversidad, Buenos Aires, Argentina
[2]Communication advisor, Arequipa Perú; carlos_gutierrez_v@hotmail.com

Abstract

At present, llamas and alpacas play an important role in the food security of populations living in high Andean areas and they also provide fibre, leather and meat. This alternative livestock is extremely important not only in arid regions of Argentina, Bolivia, Chile, Ecuador and Peru, but also in other countries that have discovered their economic potential, such as Australia, New Zealand, North America and Canada. In South America, production management is not efficient and does not generate income; they are subsistence economies. In the new globalised world it is relevant to achieve highly efficient livestock production to help territorial development without loss of identity. To such an end, we must take into account the vast experience and the great deal of information available for the last 20 years as well as the poor access to such information by the Andean communities. To begin to address these limitations, the International Agricultural Development Fund, the Biodiversity Foundation and other institutions have organised, since 2003, three editions of the 'World Congress on Camelids', with the next one to be held in Chile (12 November). Its aim is to achieve coexistence and communication between the different sectors so that there is consistency between economic, social, human and cultural issues.

The next stage will be to build a 2.0 website where all actors can actively interact through chat, blogs, social networks and forums, among other things. It will also provide a library including digitised material (Wiki), and allow for on-line training and the fostering of successful rural innovations. The project goal is to build, systematise and manage the day-to-day experience and technical knowledge in order to distribute such collective information (scaling up) trying to develop an alternative livestock production so as to achieve a productive and sustainable world in which camelid breeders are included.

Keywords: sustainable economy, technical knowledge, website

Background

In the past, the South American camelid livestock represented the wealth of the Andean region, providing products essential to the livelihoods of their communities. Llamas, alpacas, vicuñas and guanacos are originally from Argentina, Bolivia, Chile, Ecuador and Peru. At present, llamas and alpacas play an important role in the food security of the high Andean populations, since their meat is a rich source of protein. Llamas, in particular, are a means for loading and transport, and both animals provide fibre and leather for clothing and shelter, and their dung is utilised as fuel and fertiliser in some cases. Their importance as a livestock alternative in arid and semi-arid regions is increasing. This trend also exists in countries in which these species were not present some years ago and who now benefit from their potential economic productivity. These countries are Australia, New Zealand, United States of America, Canada, and on a smaller scale, France, UK, Germany, Italy, Spain, and Israel, among others.

In South America, 90% of the alpacas and 100% of the llamas are in the hands of rural communities of the Andean region. Both are highly efficient in the use of fragile natural grasslands growing in very labile soil, and are well adapted to rugged terrain and adverse weather conditions, characteristics which make this region inhospitable. Both species play a relevant role in habitat conservation and preservation of water sources, and their sustainable use is a source of income and supplies for rural communities.

It is important to mention that these regions are not highly efficient in terms of productive management. Therefore, production activities do not provide income for these communities, which presently function on the level of a subsistence economy. If this scenario is contrasted with the significant progress in other countries that have adopted this production alternative, the latter outperforms the work of centuries of these traditional communities, neglecting and poorly positioning their productivity in the market. It is of vital importance to achieve the sustainable use of the highly efficient camelid resource allowing territorial development with cultural identity, providing a holistic value to the financial, cultural, social, human, physical and natural assets of this region.

The challenge is to encourage the effective and efficient development of a livestock complex, by motivating the Andean communities to convert the productivity of llama and alpaca livestock into a valid alternative with benefits in line with the scale of production. Communities are able to learn, understand and develop better production. To achieve this, they must combine their ability (their daily learning activity) with knowledge (the acquired and systematised experiences) generating and enhancing positive experiences that allow the development of the communities, fostering the exchange of ideas for the decision-making process.

In broad terms, we must aspire to: (1) develop a native, sustainable and original livestock alternative, able to incorporate new areas (at present marginally productive) into the current livestock production regions; and, (2) offer to the market products and services – with the emphasis on their intrinsic values – arising from the sustainable use of a renewable resource. Thus, the sustainable use of camelids represents, on the one hand, an undeniable opportunity for the economic development of the high Andean communities by ensuring food security, poverty alleviation and climate change mitigation, and, on the other, a production alternative that would set in motion extensive productive regions which so far do not have an efficient production target.

Some twenty years ago, the International Fund for Agricultural Development (IFAD) supported the development of numerous initiatives involving camelid production and conservation for rural development in Bolivia and Peru. Some of these initiatives were under projects and programmes with a specific thematic approach, such as UNEPCA, PPC and PRORECA in Bolivia, others under regional projects that support camelid conservation, such as FEAS, MARENASS and CORREDOR CUSCO-PUNO, in Peru, and PROMARENA, in Bolivia.

In 2003, IFAD, and later on Fundación Biodiversidad as well as other institutions, organised in Bolivia, Ecuador and Argentina three editions of the World Congress on Camelids, whose 6th meeting will take place in November 2012, in Chile. These events were special in that they gathered together all actors in the productive complex – livestock production, processing and marketing of raw materials and products and scientific research – and achieved coexistence and communication among different sectors on significant economic, social, human and cultural interests.

What should we do with the information and expertise acquired during these last 20 years? Collective knowledge on resource management and camelid production should, at least, be carefully preserved, and access to this information for Andean communities should also be promoted and be easily available. In order to provide continuity to this process, the next stage should be to build a 2.0 Website, where what matters is the interaction between the actors and not the technology itself or its location. The primary objective is to establish a meeting point (chat, blog, forum, articles, documents, useful links, news, photos, videos, promotion and dissemination of exhibitions and events, geo-positioning projects and herds, and platform for e-learning, tutorials, contacts, supply and demand of goods and services) between farmers, scientists, manufacturers and traders, to enable continuity and expansion (scaling up) of commercial and human relations (social network) of the camelid complex. The secondary objective is to implement a library that includes digital documents and publications containing knowledge and experiences demanded by the various stakeholders. Users would therefore create, modify or delete shared texts. The aim is to promote the creation of collaborative encyclopaedias, similar to Wikipedia or other applications suitable to the coordination of information and actions, or the sharing of knowledge and texts within groups. It is important to point out that it is not about searching new structures, but reaching synergies with existing social networks and information stores and establishing universal access, sustainability and resource savings.

To achieve full participation and generate an inclusive communication and information system, it is necessary to bear in mind the points described in the following sections.

Knowledge gap and digital gap

Schematically, the 'knowledge gap' refers to the differences in access to knowledge between individuals, social groups, regions or countries, due to differences in age, economic or cultural reasons. The 'digital gap' refers to differences in terms of knowledge or access opportunities to the info-technology available to end users or regular non-professional computer users.

Modern society is characterised by benefiting from the era of knowledge, but this knowledge, which constitutes an intangible asset, is also a means of exclusion and discrimination, as financially strong societies have easier access to education and information. In this context, poor social groups in developing countries are even more excluded. This difference in the level of access to knowledge enlarges the gap between first world and third world countries.

Each time a new communication technology appears, it is exclusive. In the long run, end users have easier access to them: printing, radio, television, cellphone, Internet. Once it was assumed that the Internet depended on engineers, machines and was only for big cities, and then an evolution took place. Now, it is not important where information is stored nor whether the servers are able to accumulate such an amount of information. With Web 2.0. technology (i.e. Wikipedia and YouTube), and virtual spaces, and the programming skills of young people around the world, the point now is what should we do with the social assets of information and communication?

Human-machine binomial

There is in fact a portion of the most advanced info-technology that is wrapped in appropriate user formats both as instruments and applications, as well as services. This comes almost immediately and in waves to the public. When we refer to info-technology we mean 'Information Technology and Communication (ICT)', which is integrated into the lives of tens or hundreds of millions of end users, not technically skilled but laypeople: multifunctional mobile terminals, personal computers, PDAs, webpages, websites or webservers, blogs, digital cameras, game consoles, MP3 and MP4 players, DVD players/recorders, Blue ray, GPS navigation, e-mail, various Internet services (instant messaging, cellphone provision) cellphones with GPS and Internet, and other services linked to social networks, multimedia and many other different instruments.

These daily massive technological developments make the human-machine binomial one overwhelming reality, rather forced, chosen or rejected, depending on the context. Its behaviour and social consequences are, if analysed from different angles, a multidisciplinary field, enough to justify the need for extending or scaling-up its use to services that include 'info-poor' end users in the most remote places of our society.

In Latin America, there is significant knowledge capital regarding South American camelids: producers and breeders, scientific institutions, academic institutions, state organisations, business associations, cooperatives, among many others. These are the living memory or warehouse (library) of knowledge. This also appears to be the case in more technologically developed countries with new breeders and institutions such as Australia, New Zealand, United States of America, Canada, etc.

Digital alphabetisation

Digital alphabetisation is the development of the capacity of human-machine coupling, where the human is the one that uses the capabilities of the machines, has the criteria to do so, and expands its capabilities and actions. Having said this, the aim is to manage the 'intelligent capacities' of the machine, rather than managing the machine itself. Human logical relations should understand the relations of the digital universe. It is a fact that this does not occur in rural areas, much less in the Andean region. Therefore, if we intend to generate an exchange of knowledge and know-how among farmers, who are at the same time depositaries of traditional knowledge and petitioners of new technologies for their production system, they must have the ability and opportunity to receive digital alphabetisation.

The current inclusive experiences rely on improving digital literacy and investment in capacity-building and training activities, from personnel in charge of Internet booths in rural areas to those interested in using ICTs for their own progress. New technologies no longer depend on a reference point in which information is stored, nor on a certain number of operators to get the work done, but on keeping information *online*, through means like youtube or Wikipedia. Thus, improving digital literacy through cybertechnology should be fundamental to any proposal for including rural individuals in the market.

The social organisation of Andean communities relies on mutual benefit, trust, civic engagement and reciprocity standards. If we want these communities to generate an 'informational capital' that allows them to improve their quality of life, their participation through ICT means will consititute a strategic challenge that requires social organisations to articulate alternative cooperation networks and collective action.

A process of empowerment of ICTs by social actors should take place, directing the educational use of chats, forums and blogs. This should lead to a link between technological empowerment and improvement in their quality of live. ICT plays an increasingly important role in local development due to its ability to mobilise material resources, information and knowledge. The aim is to use the Internet for thinking and encouraging a dialogue that should result in collective knowledge, in this case regarding camelids, and not only as a means for transmitting information.

In times when culture itself has become a resource for economic development, it is necessary to move to a social appropriation of ICTs in order to deploy behavioural strength and creative capacity relying on the expertise of individuals and social actors. In order to reduce information and literacy gaps, and to attain our main objective, it will be necessary to work together with existing projects in the five countries of interest.

The 2.0 Web: a social and creative revolution

The most significant innovation of 2.0 Web is its capacity to convert a site into a platform for publication and production of contents and applications by any user. 2.0 Web is an attitude, not just a technology. It is the ability to assimilate a democratic and communicative use of the web, by exploiting the capacity for dialogue. It includes applications that generate collaboration, and services that replace desktop applications. Many say that the Internet has been reinvented.

Technologies such as blogs (*weblogs* for some Spanish-speaking users) are just a very simple mechanism for creating articles or posts in chronological order, usually short, that an individual performs on the Internet for anyone willing to read it, and enable readers to engage in a conversation with the blogger (the person who created the blog) via a feedback system through the designed posts.

Other means include wikis (the basis of the famous online encyclopedia Wikipedia), podcasts (a kind of audio blog), RSS technology (Real Simple Syndication) that allows contents to be published on the Internet from one to many (known as syndication in the publishing world), in an extremely simple way, using a system of subscription channels or social tagging and bookmarking.

As mentioned above, the most important technological innovation of 2.0 Web is to turn a website into a platform for producing and publishing contents and applications, that could be done by any user in an extremely simple way and with little technical knowledge. This concept has allowed the explosion of social applications, creativity and innovation witnessed in recent years.

Before the success of the Web, the circle of people that an individual could attract to share views and opinions was limited to physical and temporal space, limiting significantly social interaction (in fact the sociological concept of network environment is limited to two hundred contacts).

With the empowerment brought by the Internet, especially 2.0 Web, through so-called social networks, the limitations of physical space for social interactions disappears and the temporal space is much more manageable due to the possibility of asynchronous communication. Today, six out of ten most visited sites on the web are social network sites (Facebook, Twitter, hi5 and others).

To understand the implications of 2.0 Web, here are some examples of the latest Internet events and evolutions.
- Social networking
  Social networking is part of the global knowledge infrastructure. Photographs, videos, texts, articles or written comments participate in the creation of a free media, where we enter personal experiences that may serve as a reference for other social network members with whom we may share affinities and interests. Only collective action networks can deploy the necessary capacity for mobilisation and extension of social relations required for creative development in new digital technologies. These networks enhance the communication skills of its members, in space as well as in time, and multiply its effect when shared simultaneously with other networks. Each member is an active, simultaneous and permanent actor in as many proposals as desired, moving from private to public space and vice versa, without interruption. In this bidirectional communication scenario any action taken by a member of a social network could generate the creation of a new thought.

- Wikipedia
  Wikipedia, self-defined as a collaborative effort of volunteers to create a free encyclopedia in many languages, free and accessible to all. Based on wiki technology, it enables items to be written, reviewed and requested, and allows modification of a vast majority of items by anyone with access through a web browser. Presently, it depends on the non-profit Wikimedia Foundation. Since its conception, it has become one of the ten most visited websites in the world.
- YouTube
  YouTube is a website that allows users to share digital videos over the Internet. It is very popular due to the possibility of hosting personal videos easily. YouTube hosts a variety of clips from movies, TV shows, music videos and home videos (although uploading videos with copyright is banned, this material exists in abundance). Links to YouTube videos can also be placed on blogs and personal web sites using APIs or embedding some HTML code.
- Copyleft
  Copyleft, or permitted copies, includes a group of intellectual property rights characterised by removing restrictions on distribution or modification provided that the derivative work remains with the same IP scheme as the original.

The use of ICTs to link researchers, educators, extensionists and producer groups to each other and to sources of global information, or any other source, and their relationship to communication for development provides room for many different means and approaches: popular media, traditional social groups, direct interpersonal communication, rural radio for community development, video and multimedia modules for farmer training.

Through these communication resources, or even through other basic means, it is imperative that communication occurs to provide the necessary skills to interested users in order to progress, changing bad or assessing best practices and behaviors, and to achieve the sustainable development of each of the various activities.

Finally, this proposal includes a third stage that considers encouraging various successful rural innovations, replicable in other regions or communities. This will be implemented through contests or competitions between innovators, providing incentives to other farmers, and disseminating innovative practices or experiences. The aim is to promote the best livestock animal handling and improve processing of raw materials into best quality products from rural and urban families. In addition, the selected best alternatives will be rewarded with cash incentives. The steps for achieving this stage are as follows: (1) exploration of effective and efficient alternatives, and awareness of its actors to participate; (2) call to compete judged by a panel composed by the same participants of the same call; (3) incentives for the best proposals; (4) dissemination of technologies; and, (5) emulation winning alternatives. The winning alternatives will be seconded for a considerable time, in order to strengthen their weakest points, and to learn more about the factors of their success.

Conclusions

- The aim is to build an easy and fast access interactive web that can be accessed by people living in areas not served with broadband. Information must flow quickly to and from all actors, crossing, linking, affirming, rejecting, criticising, showing contents. This relationship of 'interconnected social capital' will help to generate and scaling-up knowledge.
- This tool is intended to be a communication bridge, compiling available information for farmers, artisans, traders, manufacturers, scientists, students and stakeholders, allowing them

to determine, build knowledge and profit from eventual relationships around the world of camelids.
- Priority will be given to information access to end users, building a 2.0 Web serving as a platform for improving digital literacy in those places where the digital gap still exists.
- With the current technology, the location of the site operator is not important, although this person must have a high level of information, understanding, tolerance, non-globalised universality, etc. The operator should be able to: (1) make a selection of copyright and copyleft documents, to allow its publication and copy from the Internet, or only for reading; (2) look for links treating issues related or associated with camelids; (3) guide users to specialised blogs and bloggers; (4) convene forums and dialogues; (5) build specialised communication within social networks, Twitter, Faceboock, Hi5, among others; (6) respond, locate, order and display information on camelids; and, (7) to have skills to assess qualitative and quantitative topics of projects involving camelids, and in particular, to maintain a close relationship between humans and camelids. Essentially, to sustain the existing interaction between camelids and breeders, farmers, traders, buyers, manufacturers and researchers, with a sense of humour and positiveness.
- Encourage successful ventures, support and strengthen them as well, to learn from them and systematise the means of their success, to promote replication and scaling-up.

Summarising, the fundamental aim is to build, systemise and manage knowledge collectively in order to distribute and socialise it in scale (scaling-up), with the final aim of developing a sustainable productive and inclusive alternative such as camelid conservation.

IFAD's continued involvement, with simple tools, improving the quality of life of the various actors, in particular rural communities, and the implementation of the above-described activities, will keep 'alive' information and communication between the various editions of the World Congress on Camelids, and serve to conserve South American camelids and their universe.

Preliminary study of body measurements on alpacas in northern Italy

A. Tamburini[1], A. Briganti[2], A. Giorgi[3] and A. Sandrucci[1]
[1]Department of Animal Science, University of Milan, via Celoria 2, 20133 Milan, Italy;
alberto.tamburini@unimi.it
[2]ITALPACA Association, via Vecchia di San Gervasio 47, 56025 Montecastello (PI), Italy
[3]GeSDiMont, University of Milan, via Celoria 2, 20133 Milan, Italy

Abstract

The study involved 6 farms in northern Italy and 94 alpacas of the Huacaya type. Eighteen males and 45 females were measured for height at the withers, heart girth and hip width. In each farm a questionnaire was filled out to gather information on farm characteristics and feeding strategies. The farms had on average 5.8±5.1 ha of pasture land and reared 15.7±6.4 alpacas. On average, each farm had 6.0±2.8 males. During winter or on rainy days, alpacas received on average 1.4±0.4 kg/d of commercial hay and 0.2±0.4 kg/d concentrate. The size of the farms was very small, and the alpaca business was not the primary source of income. The alpacas were 4.1±3.2 years old (female: 4.4±3.0, male: 3.5±3.6). On average, the height at the withers was 85.6±8.1 cm, the heart girth was 93.0±14.6 cm and hip width was 24.0±3.7 cm. Some regression analyses were performed: heart girth (cm) = 1.3795 wither height (cm) – 24.956 (r^2=0.57); hip width (cm) = 0.3392 wither height (cm) – 4.9761 (r^2=0.54); heart girth (cm) = 3.0198 hip width (cm) + 20.451 (r^2=0.58). In conclusion, there are good opportunities to improve alpaca farming management in northern Italy, particularly in terms of feeding and environmental factors affecting alpaca growth.

Keywords: alpaca, huacaya, body measurements, farm management

Introduction

More than 95% of the Peruvian camelid population is located in the Andes (Cervantes et al., 2010). In Italy there are fewer than 50 alpaca farms affiliated to the Italian Official Breeder Association (Italpaca, 2010); they are very small-size farms, generally distributed in marginal areas. No data are available on farm management or on body measurements of Italian Alpacas.

The aim of the study was to obtain information about alpaca farm management in northern Italy and to collect body measurements of animals.

Materials and methods

A total of 6 farms were chosen in northern Italy from the Italpaca farmer list. Each farm was visited once in the morning between June and September 2009. During farm visits a questionnaire was filled out to collect information about number and sex of animals, housing, barn and paddock design, feeding system, concentrate use, pasture areas, diseases and farm labour.

Body measurements were performed on the whole herd and included wither height (WH), heart girth (HG) and hip width (HW). Pregnancy effect was not considered. Regression analyses were performed between body measurement parameters. Coat colour of all animals was recorded by digital pictures, and clustered as suggested by Italpaca (2010).

Results and discussion

Farm size was always very small, and alpaca farm business was not a primary source of income. All the farmers referred to difficulties in fibre marketing.

The main results for the number of animals in farms are summarised in Table 1. On average, each farm reared 15.7±6.4 alpacas: 6.0 males, 2.7 primiparous females, 1.3 secondiparous females, 3.0 multiparous females and 5.7 young females. Fifty percent of the farms reared other animals, in particular llamas, angora and mohair goats: this result suggests that many small farms maintain these animals for agritourism reasons.

Farms had on average 5.8±5.1 ha of pasture land. During the grazing season all the animals were maintained on natural pastures or in olive fields. During winter or on rainy days, in all the farms alpacas received on average 1.4±0.4 kg/d of commercial hay and 0.2±0.4 kg/d of commercial concentrate. Two farms used customised concentrates and the other 4 farms used a commercial concentrate for sheep.

No farms had any kind of agricultural machine. Three farms had barns for the winter season, but animals were always allowed to graze. No animal diseases were recorded.

Body measurements were performed on 45 females and 18 males: females showed on average (Table 2) larger size than males, but an age effect could not be excluded. In fact males were

Table 1. Distribution of animals in the farms (n=6).

	Primiparous	Secondiparous	Multiparous	Young females	Males	Total	Others[1]
Mean	2.7	1.3	3.0	5.7	6.0	15.7	34.0
SD	2.1	0.6	1.4	1.2	2.8	6.3	40.7

[1] Other small ruminants in farm (llama, angora, mohair).

Table 2. Body measurements of alpacas on the monitored farms.

		Wither height (cm)	Heart girth (cm)	Hip width (cm)	Age (years)
All animals	mean	85.6	93.0	24.0	4.1
(n=63)	SD	8.1	14.6	3.7	3.2
	minimum	62	40	14	
	maximum	100	125	30	
Females	mean	86.5	94.2	24.5	4.4
(n=45)	SD	7.1	14.0	3.0	3.0
	minimum	66	40	15	
	maximum	100	125	30	
Males	mean	83.4	89.9	22.9	3.5
(n=18)	SD	10.2	16.2	5.0	3.6
	minimum	62	50	14	
	maximum	94	123	28	

younger than females (3.5±3.6 vs 4.4±3.0, respectively). Wither height was 86.5±7.1 cm for females and 83.4±10.2 for males.

Some regression equations were performed, to study gender effect.

Regression between HG and WH did not show any gender effect (Figure 1):

Males	heart girth (cm) = 1.2685 wither height (cm) − 15.822	(n=18, r^2=0.64)	(1)
Females	heart girth (cm) = 1.4387 wither height (cm) − 30.201	(n=45, r^2=0.53)	(2)
All animals	heart girth (cm) = 1.3795 wither height (cm) − 24.956	(n=63, r^2=0.57)	(3)

Regression between HP and WH showed different coefficients in males and females (Figure 2); in particular females showed a lower ability than males to increase hip width as wither height grows.

Regression equations were:

Males	hip width (cm) = 0.4443 wither height (cm) − 14.119	(n=18, r^2=0.83)	(4)
Females	hip width (cm) = 0.2459 wither height (cm) + 3.2021	(n=45, r^2=0.33)	(5)
All animals	hip width (cm) = 0.3392 wither height (cm) − 4.9761	(n=63, r^2=0.54)	(6)

Regression between HG and HP (Figure 3) did not show different coefficients in males and females. Regression equations were:

Males	heart girth (cm) = 2.6753 hip width (cm) + 28.617	(n=18, r^2=0.67)	(7)
Females	heart girth (cm) = 3.3722 hip width (cm) + 11.727	(n=45, r^2=0.53)	(8)
All animals	heart girth (cm) = 3.0198 hip width (cm) + 20.451	(n=63, r^2=0.58)	(9)

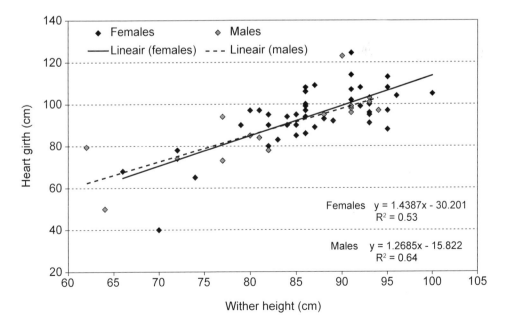

Figure 1. Regression between heart girth and wither height for females and males.

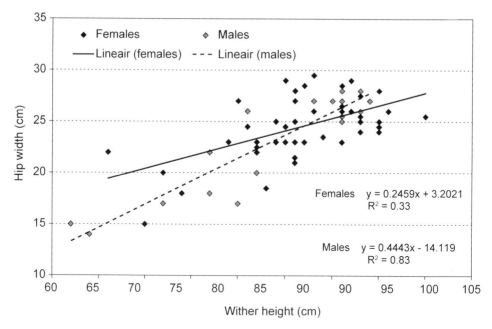

Figure 2. Regression between hip width and wither height for females and males.

For coat colours, alpacas were clustered in 3 groups (white, red, black). The most frequent coat colour was red (69% of total), particularly for females. Body measurements showed only small differences among the 3 colour groups (Table 3).

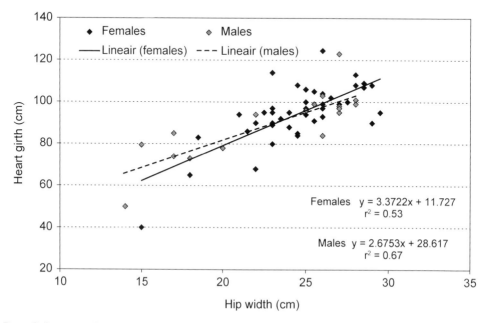

Figure 3. Regression between heart girth and hip width for females and males.

Table 3. Body measurements of alpacas clustered for coat colours.

		n	Wither height (cm)		Heart girth (cm)		Hip width (cm)	
			mean	SD	mean	SD	mean	SD
Females	white	8	87.0	6.6	95.4	15.0	25.1	4.0
	black	6	85.3	4.1	98.7	10.4	26.1	2.3
	red	31	86.6	7.7	93.1	14.6	24.0	2.8
Males	white	4	84.5	9.6	87.8	13.7	23.0	5.4
	black	2	77.5	21.9	91.3	16.6	20.5	7.8
	red	11	84.0	9.1	90.5	18.2	23.3	4.8

Conclusions

Body measurements could be a useful tool for alpaca farmers to monitor growth rate especially in young animals. Preliminary results of this study show some significant correlations between wither height, heart girth and hip width in alpaca. On the other hand there are good opportunities to improve alpaca management in northern Italy, particularly in terms of feeding and environmental factors that can affect alpaca growth.

Acknowledgements

The authors are grateful to Dr. Stefano Fenotti for collecting data and to all farmers for collaboration.

References

Cervantes I., M.A. Pérez-Cabal, R. Morante, A. Burgos, C. Salgado, B. Nieto, F. Goyache and J.P. Gutiérrez, 2010. Genetic parameters and relationships between fibre and type traits in two breeds of Peruvian alpacas. Small Ruminant Research, 88: 6-11.

Italpaca, Associazione Italiana Allevatori Alpaca, 2010. Available at: http://www.italpaca.com. Accessed September 2010.

Health

Orthopaedic problems in llamas and alpacas: clinical and radiological aspects

I. Gunsser
LAREU, Munich, Germany; Ilona.Gunsser@t-online.de

Abstract

Llamas and alpacas show only reduced signs of pain in cases of orthopaedic problems. For that reason early indications of orthopaedic problems might be overlooked. In some cases orthopaedic problems could be inherited. X-rays of 195 llamas and alpacas were analysed for problems in the region of the spine and the legs. The radiographs were taken as a diagnostic aid in patients or were made for scientific purposes from carcasses or parts of it. The results revealed problems which are similar in other domestic species, such as fractures, joint infection, osteoporosis, rickets, patella luxation, spondylosis, carpus angular deformity, and malformation of the elbow. More specific to the anatomy of camelids are the subluxation of the vertebra of the cervical spine and problems with the pastern. Examples of these findings will be presented and discussed.

Keywords: legs, spine, radiograph, welfare

Introduction

More and more alpacas and llamas are kept as companion animals, or in the case of alpacas mainly for fibre production. Since owners are quite frequently unaware of the fact that South American camelids show only reduced signs of pain, early indications of orthopaedic problems or other injuries might be overlooked. It should be noted, however, that some of these orthopaedic problems might be inherited. The knowledge of possible orthopaedic problems and their clinical aspects is therefore quite important for animal welfare reasons.

Materials and methods

X-rays of 195 alpacas and llamas were taken and analysed for problems in the region of the spine and the legs. Many of these radiographs had been taken as a diagnostic aid in patients showing abnormal position in these anatomical regions or other problems like swellings or lameness. The majority, however, had been made for scientific purposes from carcasses or parts thereof. The information about clinical signs like swelling or lameness had been collected by the author or had been reported by the owner.

Results and discussion

The radiographs revealed problems which are similar in other domestic species, such as leg fractures in the region of the long bones. Some animal had joint infections and some had arthrosis. Other findings were signs of osteoporosis or rickets. Further problems were caused by malformation of the elbow or the incorrect position of the patella. Knock-knees were found rather frequently, as a result of the incongruent closure of the growth plate of the forearm. Occasionally signs of spondylosis of the spine were seen (Gunsser *et al.*, 1997. See also Fowler, 1998). More specific to the anatomy of camelids are the subluxation of the vertebra of the cervical spine and problems with the pastern (Gunsser, 2009).

Subluxation of the vertebra of the cervical spine

South American camelids (SACs) have a rather long neck. The length of the neck corresponds to the length of the back, excluding the pelvis. The cervical spine has 7 rather long vertebras which articulate with their vertebral processes at both sides with the vertebras following behind. In addition, strong ligaments which insert at the dorsal part of the vertebras connect the cervical spine to the thoracic spine. A muscular system along the neck provides balance while walking and allows very flexible movements to each side of the body. In case of accidents these bony articulations of the vertebra may fracture. If the articulation of one side has fractured, the result is a horizontal subluxation.

In some SACs we have seen a malformation of the vertebral articulation (see Figure 1). The length of the articulation was not symmetric and only one side had good contact with the vertebral process. In these cases the horizontal subluxation can easily occur. In both of these cases the clinical signs are a flexure to one side, mostly in the mid-region of the neck (Figure 2). Neurological signs are not obvious, or hardly noticeable. If both articulations had fractured, the result would be a subluxation or luxation in all directions with greater neurological damage. Presumably the malformation of the vertebra is a heritable defect.

Problems with the pastern joint

Since SAC belong to the tylopoda, the structure of the foot differs from that of other domestic animals like ruminants. The foot of a ruminant touches the ground with the third phalanx of each of the two toes. Each toe is covered with a claw. The foot of the camelid, on the other hand, touches the ground with the second and third phalanx of the toes. The toes are equipped with a digital cushion. Walking on two toes requires well-balanced, symmetrical toes. The toes articulate in the pastern joint with the metacarpal - /metatarsal bone. In case of an incongruence in one of these bones, the length and direction of the toes will be different (see Figure 3). The result is a strain on the articulation and tendons, leading to chronic irritation and arthrosis. In fact these malformations are rarely diagnosed in new-born alpacas and llamas. The reason for

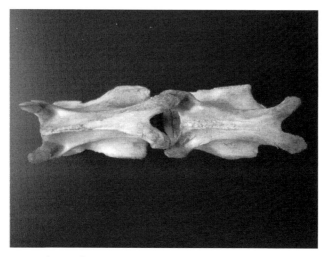

Figure 1. Preparation of the 4th and 5th cervical vertebra. A malformation of the vertebra (asymmetric articulation process), alpaca, 1 year.

Figure 2. Alpaca, subluxation of a vertebra of the neck.

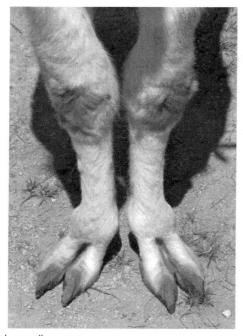

Figure 3. Abnormal form of the toe, llama.

this is the relatively small foot of these animals. During the growth period the incongruence of the toes becomes more pronounced. When this is finally noticed, the owner might misinterpret the incongruence as resulting from an accident with fracture.

Another finding in the pastern joint is an instability of the pastern joint. In such cases the angulation of the phalanx of one of the toes is less than about 60 degrees to the ground. Within relatively little time (some months) the pastern joint may get weaker until it touches the ground (hyperextension of the fetlock, see Figure 4). This leads to an overstrain of the tendons and articulation, resulting in pain caused by arthrosis and chronic irritation of the tendons (see Figure 5). Most affected SACs show these problems bilaterally, starting at about 2 years. A hyperextension may also arise from unilateral over-stressing of the fetlock, for example after amputation of one leg. The reason for a bilateral hyperextension is unclear. Since such problems have been found within near relatives, an inherited component may be likely. It has been argued in the past that early castration of males may also result in hyperextension of the fetlock. However, this opinion cannot be maintained since such problems also exist in females.

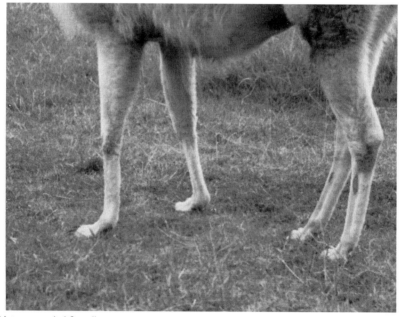

Figure 4. Hyperextended foot, llama.

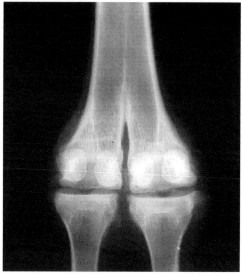

Figure 5. Radiograph arthrosis, hyperextended fetlock, llama.

Conclusion

Orthopaedic problems in llamas and alpacas do not differ much from the problems known in other domestic animals. However, the specific anatomy in the region of the neck and foot can lead to characteristic problems in SACs. For the purposes of animal welfare it is important to inform owners about possible pain-inducing orthopaedic problems in their animals. Up to now it is not certain which of the orthopaedic problems are inheritable. Nevertheless it is important to check breeding males for orthopaedic abnormalities using clinical and radiological techniques.

References

Fowler Murray E., 1998. Medicine and surgery of South American Camelids, 2. Aufl. Iowa State University Press, Ames, IA, USA.

Gunsser I., H. Wiesner and T. Hänichen, 1997. Krankheiten (Diseases). In: M. Gauly (ed.), Neuweltkameliden. Parey Verlag, Berlin, Gemany, pp. 120-154.

Gunsser I., 2009. Problems in keeping llamas and alpacas in Germany, 2nd Conference of ISOCARD, Djerba, Tunisia, March 12-14.

Diarrheagenic *Escherichia coli* strains isolated from neonatal Peruvian alpacas (*Vicugna pacos*) with diarrhea

D. Cid[1], C. Martín-Espada[1], L. Maturrano[2], A. García[1], L. Luna[2] and R. Rosadio[2]
[1]*Departamento de Sanidad Animal, Facultad de Veterinaria, Universidad Complutense de Madrid, Avda. Puerta de Hierro s/n, 28040 Madrid, Spain; lcid@vet.ucm.es*
[2]*Unidad de Biología y Genética Molecular, Departamento de Salud Animal, Facultad de Medicina Veterinaria, Universidad Nacional Mayor de San Marcos, Av. Circunvalacion 2800 San Borja, Lima, Peru*

Abstract

Neonatal diarrhea is a complex infectious disease caused by the interaction of adverse environmental conditions, deficient immune status of neonates and infection of enteropathogens. A variety of microorganisms can cause diarrhea in alpaca neonates, including bacteria, viruses and parasites. In the 1980s, isolation of *Escherichia coli* strains capable of inducing fluid accumulation in the intestinal loop assay in alpaca and llama neonates was described. However, virulence factors of *E. coli* associated with neonatal diarrhea in South American camelids (SAC) have been poorly characterised. In this study, 94 *E. coli* isolates from 94 crias of alpaca (1 per animal) with diarrhea were analysed for the presence of virulence factor genes of diarrheagenic *E. coli* by PCR. The investigated genes were verotoxin (*vt1 and vt2*), intimin (*eae*), and bundle-forming pili (*bfp*) genes. The animals belonged to six different herds of the Andean region of Cuzco (Peru) and were between two and nine weeks old. Two categories of diarrheagenic *E. coli* strains were detected: Enteropathogenic *E. coli* (EPEC), and Verotoxin-producing *E. coli* (VTEC) strains. Out of the 94 analysed strains 2.13% (2/94) were EPEC (*eae* and *bfp*), 10.64% (10/94) were atypical EPEC (*eae* but *bfp* negative), and 11.7% (11/94) were VTEC. Only 3 of the VTEC strains (3/11) were *eae* positive: one *eae vt2* and two *eae vt1*. All of the other VTEC strains (8/11) were only *vt1*. Infection with diarrheagenic *E. coli* strains was detected in all the herds. The results indicate that these pathogenic strains are circulating among domestic SAC, probably at high rates, in the Andean region of Peru. Therefore, EPEC and VTEC strains may be a leading cause of diarrhea in these animals. In addition, these pathogens are a matter for public health. Preventive measures based on vaccination and improvement of hygiene conditions are essential to reduce the impact of neonatal diarrhea in SAC health and mortality in the Andean region of Peru.

Keywords: enteropathogenic, verotoxigenic, *Escherichia coli*, diarrhea, alpaca

Introduction

South American camelids (SAC) production is the most important means of subsistence for the Andean communities of Peru living at high altitude, most of them in conditions of poverty or extreme poverty. Alpaca, *Vicugna pacos*, is one of the two domestic species of South American camelids (SAC) which supplies high-quality products such as fibre and meat. Neonatal mortality is one of the most limiting factors of alpaca production in these rural communities. High rates of neonatal mortality raising 50% or 80% of the born cria have been reported (Bustinza *et al.*, 1998; Fernández Baca, 2005). The major causes of neonatal mortality in these conditions are infectious diseases, starvation, poor hygiene conditions and inadequate handling.

Among infectious diseases, neonatal diarrhea is one of the leading causes of neonatal mortality of alpaca. Neonatal diarrhea is a complex infectious disease caused by the interaction of adverse environmental conditions, deficient immune status of neonates and infection by enteropathogens. A variety of microorganisms can cause diarrhea in neonates including bacteria, viruses and parasites (reviewed in Whitehead and Anderson, 2006; Martin Espada *et al*., 2010).

Escherichia coli are normal habitants of the intestines of mammals, however some pathogenic strains are capable of inducing diarrhea in humans and in domestic animals. At least six different diarrheagenic categories are recognised in humans according to virulence factors, serotype and mechanisms to induce diarrhea (Nataro and Kaper, 1998; Kaper *et al*., 2004). Enteropathogenic *E. coli* (EPEC) and Verotoxin-producing *E. coli* (VTEC) are the categories more frequently associated with neonatal diarrhea in animals.

EPEC are defined as *E. coli* possessing intimin gene (*eae*) but lacking verotoxin (or Shiga-like toxin) genes (Nataro and Kaper, 1998). EPEC strains cause a characteristic histological lesion in enterocytes termed attaching and effacing lesion which consists of intimate adhesion of bacteria to enterocyte membrane, destruction of brush border, and finally detachment of enterocytes. Typical EPEC possess a fimbrial adhesin called bundle-forming pili (*bfp*) encoded in an EAF plasmid (EPEC *adherence factor*) by *bfp* gene. EPEC harbouring *eae* genes but lacking *bfp* genes are termed atypical EPEC (Nataro and Kaper, 1998).

For many years typical EPEC have been recognised as a leading cause of infantile diarrhea in developing countries and were considered rare in industrialised countries (Nataro and Kaper, 1998). However, recent data suggest that atypical EPEC are more prevalent than typical EPEC in both developing and developed countries and some strains of atypical EPEC have caused diarrhea outbreaks in the United Kingdom, the United States of America (USA) and Japan (Ochoa *et al*., 2008).

VTEC strains, also called Shiga-toxin producing strains (STEC), are defined as *E. coli* strains producing verotoxins or Shiga-like toxins (Nataro and Kaper, 1998). The family of VTs contains two subgroups, VT1 (Stx1) and VT2 (Stx2) encoded by *vt1* and *vt2* genes respectively. VTEC strains may produce *vt1* or *vt2* alone, or both types. In addition to VTs, some VTEC strains also produce intimin proteins homologous to those that are produced by EPEC strains and are able to induce attaching and effacing lesions.

Enterohemorrhagic *E. coli* strains (EHEC) are a subset of VTEC strains which cause two severe diseases in humans, haemorrhagic colitis (HC) and haemolytic uremic syndrome (HUS). Most EHEC strains produce VTs and attaching and effacing lesions and belong to a limited number of serotypes, mainly O157:H7, but EHEC strains that are negative for *eae* gene are also associated with human disease (Kaper *et al*., 2004). EHEC is the most important recently emerged food-borne pathogen. Animals, mainly cattle, are the reservoir of EHEC pathogenic for humans. Food of animal origin, such as undercooked meat and unpasteurised milk, and fecal-contaminated vegetables are major vehicles for the transmission of EHEC from animals to humans (Karmali, 2004).

In animals, EPEC and VTEC strains are frequently isolated from diarrheic and healthy animals and have also been associated with diarrhea in young animals (reviewed in Mainil and Daube, 2005; Wales *et al*., 2005; Gyles, 2007). Natural infection with EPEC and VTEC producing attaching and effacing lesions have been reported in most of the domestic animal species and experimental infections have demonstrated that some of these strains are able to induce diarrhea in calves. However, only some pathotypes, determined on the basis of intimin and verotoxin types and serotype, seem to be related to diarrhea in domestic animals (Wieler *et al*., 1998; Cid *et al*., 2001; Blanco *et al*., 2004).

Little is known about the epidemiology of neonatal diarrhea in alpacas in the Andean region. In the 1980s, isolation of *E. coli* strains capable of inducing fluid accumulation in the intestinal loop assay in alpaca and llama neonates was described (Ramírez *et al.*, 1985). However, virulence factors of *E. coli* associated with neonatal diarrhea in South American Camelids (SAC) have been poorly characterised. In this study, we investigated the possible role of different diarrheagenic categories of *Escherichia coli* in alpaca neonatal diarrhea.

Material and methods

Animals

Fecal samples were obtained by rectal swab from 94 cria of alpaca with diarrhea in the Andean region of Cuzco (Peru). Animals belonged to six different herds and were between two and nine weeks old.

Bacterial strains

For the *E. coli* isolation, fecal samples were plated onto McConkey agar and the plates were incubated at 37 °C for 18 h. *E. coli* isolates were preserved at room temperature onto semisolid nutrient broth. A total of 94 *E. coli* isolates, one per animal, were analysed for the presence of virulence factor genes of diarrheagenic *E. coli*.

Detection of virulence factor genes by PCR

The investigated genes were verotoxin (*vt1* and *vt2*), intimin (*eae*), and bundle-forming pili (*bfp*) genes. The detection was carried out by a multiplex PCR. The used primers have been previously described (Blanco *et al.*, 2004; Cordeiro *et al.*, 2009). PCR reactions were performed in 20 µl reaction mixture with Multiplex PCR kit (Qiagen). Amplification of PCR products was performed by a denaturation step at 95 °C for 15 min and 30 cycles of 94 °C for 30 s, 56.5 °C for 90 s, and 72 °C for 90 s, followed by a final extension at 72 °C for 10 min in a thermal cycler (Bio-Rad). PCR products were detected on 1% agarose gels.

Results

E. coli isolates were distributed into different categories of diarrheagenic *E. coli* according to the gene detected (Table 1). Isolates positive to *eae* and *bpf* genes but negative to *vt1* or *vt2* genes were classified as EPEC; isolates positive only to *eae* gene were classified as atypical EPEC; and isolates positive to *vt1* and/or *vt2* were classified as VTEC strains.

Two categories of diarrheagenic *E. coli* strains were detected: Enteropathogenic *E. coli* (EPEC), and Verotoxin-producing *E. coli* (VTEC) strains (Table 1). Out of the 94 analysed strains 2.13% (2/94) were EPEC (*eae* and *bfp*), 10.64% (10/94) were atypical EPEC (*eae* but *bfp* negative), and 11.7% (11/94) were VTEC. Only 3 of the VTEC strains (3/11) were *eae* positive: one *eae vt2* and two *eae vt1* (Table 2). All of the other VTEC strains (8/11) were only *vt1*.

Infection with diarrheagenic *E. coli* strains was detected in all the herds (Table 3). All the herds were infected with at least one category of diarrheagenic *E. coli* strains. Infection with VTEC seems to be widespread among herds since VTEC strains were detected in all the herds. VTEC infection was detected in diarrheic animals between two and nine weeks old.

Table 1. Frequencies of diarrheagenic categories of E. coli detected in diarrheic cria of alpaca.

Categories	Genes detected	Frequency	%	Cumulative %
EPEC	eae + bfp	2	2.13	2.13
Atypical EPEC	eae	10	10.64	12.77
VTEC	vt1 and/or vt2	11	11.7	24.47
Negative	None of the analysed	71	75.53	100
Total		94	100	

Table 2. Characteristics of 11 VTEC strains isolated from cria of alpaca with diarrhea: verotoxin type and presence of eae gene.

VTEC	Frequency	%
eae+ vt1	2	18.18
eae+ vt 2	1	9.10
eae+ vt1+ vt2	0	0.00
vt1	8	72.72
vt2	0	0.00
Total	11	

Table 3. Distribution of diarrheagenic categories of E. coli isolates among herds of alpaca.

Herd	EPEC	Atypical EPEC	VTEC	Negative	Total
IPC	1	3	3	30	37
Julian	0	0	1	2	3
Korpacancha	0	3	3	9	15
Plantel	0	0	1	1	2
Rayapata	1	0	1	7	9
Santa Lucia	0	4	2	22	28

Discussion

Attaching and effacing intestinal lesions and natural infection with EPEC in cria of alpaca have been previously described in the United Kingdom (Foster et al., 2008; Twomey et al., 2008). Foster et al. (2008) detected the attaching and effacing lesion characteristics of EPEC in three crias of alpaca, two dying from diarrhea and one from septicaemia, but bacteria were not isolated. Twomey et al. (2008) described multifocal bacterial colonisation with attaching and effacing organisms similar morphologically to E. coli in one cria of alpaca with Cryptosporidiosis. In this study, we describe the detection of EPEC infection in cria of alpaca with diarrhea in the Andean region of Peru. Most of the EPEC strains detected were atypical EPEC since they lacked the bfp gene. Typical EPEC have been isolated from animals, but most of EPEC of animal origin lack the bfp gene, so they are classified as atypical EPEC (Cid et al., 2001; Blanco et al., 2004). Some EPEC strains have been associated with diarrhea in domestic animals and EPEC is one of the most prevalent diarrheagenic categories of E. coli in diarrheic ruminants. Thus, infection with the EPEC strains isolated from diarrheic cria of alpaca in this study could be associated with the disease.

Mercado *et al.* (2004) described the association of one O26:H11 VTEC strain with diarrhea in a guanaco in Argentina. The authors considered that the source of contamination with this strain was unclear since previous contamination of the environment with bovine fecal bacteria could be the source of infection for diarrheic guanaco in this case. However, they did not discount the possibility of the existence of a resident VTEC population in guanacos. VTEC strains have also been detected in diarrheic cria of alpaca in Peru (Luna, 2009).

In this study, VTEC strains were detected in a relative high frequency in diarrheic cria of alpaca in the Alto Andino region of Peru. The most frequently detected verotoxin type in these VTEC strains was VT1. These results are in agreement with the results in other domestic species. Most VTEC strains isolated from diarrheic animal are mainly *vt1* strains whereas VTEC strains producing *vt2*, or both *vt1* and *vt2*, are most frequently isolated from healthy animals (Blanco *et al.*, 2001). The VTEC strains isolated from diarrheic alpaca in this study could play a role in the etiology of neonatal diarrhea in these animal species in the Andean region of Peru.

VTEC strains have been also isolated from healthy alpacas in Europe and South America (Pritchard *et al.*, 2009; Leotta *et al.*, 2006; Cordero *et al.*, 2009). VTEC strains were detected in two alpacas in the Zoo of La Plata, Argentina, (Leotta *et al.*, 2006). Pritchard *et al.* (2009) detected O157 VTEC strains in CSA, four alpacas and nine llamas, from public amenity premises in England and Wales between 1997 and 2007. Recently, the isolation of VTEC strains at high frequency (29.3%) from healthy alpaca in the Alto Andino region of Peru has been described (Cordero *et al.*, 2009). Thus, infection with VTEC strains in alpacas may represent a risk factor for human health in this area.

The results indicate that EPEC and VTEC strains are circulating among domestic SAC, probably at high rates, in the Andean region of Peru. Therefore, EPEC and VTEC strains may be a leading cause of diarrhea in these animals. In addition, these pathogens are concerned with public health. Preventive measures based on vaccination and improving of hygienic conditions are essential to reduce the impact of neonatal diarrhea in SAC health and mortality in the Andean region of Peru.

Conclusion

EPEC and VTEC strains may be a leading cause of diarrhea in alpacas in the Andean region of Peru. In addition, these pathogens are a matter for public health.

Acknowledgements

This work has been supported by the Universidad Complutense de Madrid, Proyecto de Cooperación Internacional para el Desarrollo V convocatoria, and by the Agencia Española de Cooperación Internacional para el Desarrollo (AECID), Proyecto A/017921/08.

References

Blanco J., M. Blanco, J.E. Blanco, A. Mora, M.P. Alonso, E.A. González and M.I. Bernárdez, 2001. Epidemiology of verocytotoxigenic *Escherichia coli* (VTEC) in ruminants. In: G. Duffy, P. Garvey and A. McDowell (eds.) Verocytotoxigenic *E. coli.*. Food and Nutrition Press, Trumbull, CT, USA, pp. 113-148.

Blanco M., J.E. Blanco, A. Mora, G. Dahbi, M.P. Alonso, E.A. González, M.I. Bernárdez and J. Blanco, 2004. Serotypes, virulence genes, and intimin types of Shiga toxin (verotoxin)-producing *Escherichia coli* isolates from cattle in Spain and identification of a new intimin variant gene (eae-xi). Journal of Clinical Microbiology, 42: 645-51.

Bustinza AV, P.J. Burfening and R.L. Blackwell, 1988. Factors affecting survival in young alpacas (*Lama pacos*). Journal of Animal Science, 66: 1139-1143.

Cid D., J.A. Ruiz-Santa-Quiteria, I. Marín, R. Sanz, J.A. Orden, R. Amils and R. de la Fuente, 2001. Association between intimin (eae) and EspB gene subtypes in attaching and effacing *Escherichia coli* strains isolated from diarrhoeic lambs and goat kids. Microbiology, 147: 2341-5233.

Cordeiro A., M. Blanco, W. Huanca, A. Mora, A. Herrera, B. Puentes, E. Quina, C. Lopez, R. Mamani, J.E. Blanco and J. Blanco, 2009. Determinación de serotipos y genes de virulencia de *Escherichia coli* verotoxigénicos (ECVT), en coprocultivos de alpacas (*Vicugna pacos*) clínicamente sanas, procedentes de puna húmeda del Perú. XXI Meeting of the Latin American Association of Animal Production (ALPA). 18-23 October.

Fernández Baca S., 2005. Situación actual de los camélidos sudamericanos en Perú. Proyecto de Cooperación Técnica en apoyo de la crianza y aprovechamiento de los Camélidos Sudamericanos en la Región Andina TCP/RLA/2914. Food and Agriculture Organisation (FAO).

Foster A.P., A. Otter, A.M. Barlow, G.R. Pearson, M.J. Woodward and R.J. Higgins, 2008. Naturally occurring intestinal lesions in three alpacas (*Vicugna pacos*) caused by attaching and effacing *Escherichia coli*. Veterinary Record, 162(10): 318-320.

Gyles C.L., 2007. Shiga toxin-producing *Escherichia coli*: An overview. Journal of Animal Science, 85: E45-E62.

Kaper J.B., J.P. Nataro and H.L. Mobley, 2004. Pathogenic *Escherichia coli*. Nature reviews. Microbiology, 2: 123-140.

Karmali M.A., 2004. Shiga toxin-producing *Escherichia coli*. Molecular Biotechnology, 26: 117-122

Leotta G.A., N. Deza, J. Origlia, C. Toma, I. Chinen, E. Miliwebsky, S. Iyoda, S. Sosa-Estani and M. Rivas, 2006. Short communication. Veterinary Microbiology, 118: 151-157.

Luna L., 2009. Genotipificación de cepas de *Escherichia coli* aislados de crías de alpacas con diarreas. Thesis, Faculty of Veterinary Medicine, Univesity of San Marcos, Lima, Peru, p. 70.

Mainil J.G. and G. Daube, 2005. Verotoxigenic *Escherichia coli* from animals, humans and foods: who's who? Journal of Applied Microbiology, 98:1332-1344.

Martín Espada C., C.E. Pinto Jiménez and D. Cid Vázquez, 2010. South american camelids: health status of their cria. Revista Complutense de Ciencias Veterinarias. 4: 37-50.

Mercado E.C., S.M. Rodríguez, A.M. Elizondo, G. Marcoppido and V. Parreño, 2004. Isolation of shiga toxin-producing *Escherichia coli* from a South American camelid (*Lama guanicoe*) with diarrhea. Journal of Clinical Microbiology, 42(10): 4809-11.

Nataro P.J. and J.B. Kaper, 1998. Diarrheagenic *Escherichia coli*. Clinical Microbiology Reviews, 11: 142-201.

Ochoa T.J., F. Barletta, C. Contreras and E. Mercado, 2008. New insights into the epidemiology of enteropathogenic *Escherichia coli* infection. Transactions of the Royal Society of Tropical Medicine and Hygiene, 102: 852-856.

Pritchard G.C., R. Smith, J. Ellis-Iversen, T. Cheasty and G.A. Willshaw, 2009. Verocytotoxigenic *Escherichia coli* O157 in animals on public amenity premises in England and Wales, 1997 to 2007. Veterinary Record, 164: 545-549.

Ramírez A., R. Ellis, J. Sumar and V. Leyva, 1985. *E. coli* enteropatógena en alpacas neonatales: aislamiento de intestino delgado y su inoculación oral. In: Proceedings of the V International Conference on South American Camelids, Cusco, Peru, p. 34.

Twomey D.F., A.M. Barlow, S. Bell, R.M. Chalmers, K. Elwin, M. Giles, R.J. Higgins, G. Robinson and R.M. Stringer, 2008. Cryptosporidiosis in two alpaca (*Lama pacos*) holdings in the South-West of England. Veterinary Journal, 175: 419-422.

Wales A.D., M.J. Woodward and G.R. Pearson, 2005. Attaching-effacing bacteria in animals. Journal of Comparative Pathology, 132: 1-26.

Whitehead, C.E. and D.E. Anderson, 2006. Neonatal diarrhea in llamas and alpacas. Small Ruminant Research, 61: 207-215.

Wieler L.H., A. Schwanitz, E. Vieler, B. Busse, H. Steinrück, J.B. Kaper and G. Baljer, 1998. Virulence properties of Shiga toxin-producing *Escherichia coli* (STEC) strains of serogroup O118, a major group of STEC pathogens in calves. Journal of Clinical Microbioly, 36: 1604-1607.

Abstracts

Changes in suckling behaviour during lactation in llamas (*Lama glama*)

A. Klinkert and M. Gerken
Department of Animal Sciences, University of Göttingen, Albrecht-Thaer-Weg 3, 37075, Göttingen, German; arubel@gwdg

Abstract

Suckling behaviour plays an important role in the normal growth of progeny in mammals. There are only a few data on suckling behaviour in South American camelids (SAC), and long-term suckling behaviour is not well documented. The objective of our study was to observe the development of suckling behaviour during lactation and determine diurnal rhythms. Five female llamas and their crías were held under stable conditions (light schedule from 06:00 to 22:00 h) and videotaped with their offspring over two 24 h periods in 4 testing weeks (week 3, 10, 18 and 26 *post partum* (p.p.)). Videos were analysed for total suckling time, duration of single suckling bouts and suckling frequency, defining suckling events as a contact between the foal's mouth and the udder for more than 5 sec. Furthermore, the diurnal distribution of suckling events was examined. Results show a decreasing frequency of suckling events during lactation, reaching 57.4 ± 35.8 (mean $\pm$SD), 56.8 ± 19.7, 31.1 ± 13.6 and 24.7 ± 8.0 events per 24 h at week 3, 10, 18 and 26 p.p., respectively. The average daily suckling time showed a similar development, decreasing from 78.6 ± 37.6 min in week 3 p.p. to 69.6 ± 18.2 min in week 10 p.p. and subsequently falling to 51.6 ± 12.2 min in week 18 p.p. and 41.0 ± 9.0 min in week 26 p.p. The suckling time and suckling frequency varied strongly between individuals, ranging from 42.9 to 78.3 min/day and 25 to 63 suckling events/day across all measurement dates. Mean suckling duration/h averaged 2.8 ± 1.4 min during the day and 2.0 ± 0.7 min during the night. Thus, 26.5% of all suckling was performed during the 8 night hours and 73.5% during the light period. The highest suckling activity was observed during the last light hour (21:00-22:00), reaching 4.22 ± 4.07 min/h. Regarding the circadian rhythmic of suckling behaviour no differences could be observed during the lactation period.

Anatomical imaging of blood vessels for venipuncture in South American camelids

C. Schulz[1], K. Amort[2], M. Gauly[3], M. Kramer[2] and K. Koehler[4]

[1] *Institute of Parasitology, Justus Liebig University Giessen, Rudolf-Buchheim-Str. 2, 35392 Giessen, Germany; claudia.schulz@vetmed.uni-giessen.de*

[2] *Department of Veterinary Clinical Sciences, Clinic for Small Animals, Justus Liebig University Giessen, Frankfurter Str. 108, 35392 Giessen, Germany*

[3] *Department of Animal Science, Livestock Production Group, Georg August University Göttingen, Albrecht-Thaer-Weg 3, 37075 Göttingen, Germany*

[4] *Institute of Veterinary Pathology, Justus Liebig University Giessen, Frankfurter Str. 96, D-35392 Giessen, Germany*

Abstract

Anatomical knowledge is fundamental for good veterinary practice to ensure the health and welfare of animals under veterinary care. Correct venipuncture is an inevitable medical skill for further diagnostic examination, intravenous injection or infusion especially in sick animals. In Germany, South American camelids (SAC) are becoming increasingly popular, but are still rather uncommon patients and a neglected issue in veterinary medical education. Blood sampling of SAC is possible at different locations but preferably conducted on the jugular vein to obtain larger quantities. Due to anatomical peculiarities, such as the absence of a jugular furrow, the blood sampling from the vena jugularis is more difficult in SAC than in ruminants or horses. Presently, hardly any anatomical studies are available. The aim of the present anatomical study was to illustrate correct locations of venipuncture by colouration of the vena jugularis, arteria carotis, truncus vagosympaticus trunks and the oesophagus. Surrounding structures were prepared to improve the anatomical comprehension. Further preparations were carried out of the vena brachialis on the cranial forearm, the vena and arteria saphena as well as the nervus femoralis on the medial side of the stifle and the superficial vena coccygealis of the tail. Ear veins were also considered in preparation as they can be easily accessed to collect small amounts of blood, e.g. for the diagnosis of pregnancy or DNA analyses. Anatomical pictures will be illustrated in comparison with pictures of blood collection in living SAC to link theory with practice on the poster. In addition, to clarify anatomical, unprepared locations of superficial and subjacent structures of the SAC neck, images of the cervical region were produced by computed tomography as well as magnetic resonance tomography and compared with transverse sections of the afterwards frozen neck. The presented pictures give detailed information on possible locations for venipuncture in SAC facilitating blood collection for veterinarians and SAC owners.

Alpaca breeding and production prospects in the United States

T. Wuliji
University of Nevada, Reno, NV 89557; tumenw@yahoo.com

Abstract

The alpaca is the most important fibre-producing member of the South American camelids. Although camelids are adapted to the highland environment of the Altiplano, their introduction in some other regions of the world have been successful. Alpacas were brought in the United States prior to 1980, but were confined mostly to zoo and zoological parks. Since the first importation from Chile in 1984, purchase and breeding alpacas in the US have flourished, and since then have been distributed in every state of the US from Alaska to Arizona and California to the Carolinas, creating a sizable specialty livestock and fibre production industry. There are an estimated 300,000 camelids in North America. The alpaca owners & breeders association (AOBA) was organised in 1987 to serve and coordinate the alpaca industry developments. The AOBA provide a number of services to members and producers including an alpaca magazine, website, newsletters, investment information, scientific literatures, fibre education, alpaca shows and sales. Currently, AOBA has more than 4,000 members and registered 171,316 alpacas at the Alpaca Registry Inc (ARI) from 1986 to 2010. The Alpaca Fiber Cooperative of North America (AFCNA) is a society of more than 1000 alpaca producers who collect, process and market alpaca fibre products for the members. In 1997, the AOBA established the Alpaca Research Foundation (ARF) to support camelid research, which benefits the North American alpaca industry, primarily in the areas of alpaca health, husbandry, genetics and fibre.

Round tables

Common management denominators between South American camelids and other fibre animals

Chair: Martina Gerken (Germany)
Chairpersons: Daniel Allain (France), Renzo Morante (Peru), Susan Tellez (USA) and Julieta von Thüngen (Argentina)
Reporter: Annegret Klinkert (Germany) and Martina Gerken (Germany)

Different scientific groups are working with either South American camelids or other fibre animal species. This round table offered a platform to identify common management denominators between these species.

D. Allain (France) introduced the topic by giving an overview of fibre biology of the main fibre-producing animal such as sheep, angora and cashmere goats, South American camelids (SAC) and angora rabbits. Although there are species-specific differences in fleece composition and fibre structure, there are similar selection criteria for genetic improvement: fibre quantity (clean fleece weight) and fleece quality. The latter includes fibre diameter, fibre length, fibre distribution (standard deviation and CV, coefficient of variation), ratio between secondary and primary follicles and percentage of undesirable fibres (no kemp or coarse fibre), no medullation (in wool, mohair and cashmere) or bristles (angora rabbit). Fibre length mainly depends on the shearing intervals. Fibre colour is an important quality trait, but has no influence on fibre diameter. In sheep, goats and SAC mainly additive polygenic effect and/or major genes are acting, while in angora rabbits autosomal recessive genes are also important. Index selection represents the most efficient tool for genetic improvement of fibre quantity and quality, however, genetic correlations have to be considered as these may differ between fine fibre animal species.

The situation of fibre-producing animals for different regions of the world

J. Von Thüngen (Argentina): Argentina has quite diverse eco-regions. In the lowlands soybean production represents the main agricultural activity. About 70% of the surface consists of mainly natural grassland. The pastures of Patagonia are characterised by a co-existence of white Merino sheep, goats, llamas and wild guanacos. Apart from Angora goats, criollo goats are found with a double coat comparable to cashmere. In the Mapuche community a special criollo sheep with different colours (also spotted) named 'Linca' is used for local homespun products sold to tourists.

R. Morante (Peru): The use of farm animals varies with altitude. In the coastal area, agriculture and dairy farms dominate up to 1,500-2,000 m. At higher altitudes, sheep are kept, while alpacas and llamas are the main fibre-producing animals above 3,000 m. Problems arise when alpacas and llamas are kept together and allowed to interbreed. Alpacas are shorn twice per year. Smallholders frequently keep alpacas and sheep together to balance economic risks. Sheep are then used as capital which can be sold as occasion demands. In SAC the main income comes from fibre, but also from meat. The campesinos consume mainly alpaca and llama meat, while alpaca and sheep meat is sold to the towns. In the Peruvian textile industry, sheep wool is important to run the fibre industry 12 months a year. In addition, alpaca and llama fibre is processed and also guanaco fibre exported from Argentina. To obtain more wear-resistant textiles, alpaca fibre and wool are frequently mixed, with sheep wool used for the warp and alpaca fibre in the woof, because sheep wool has higher tensile strength than alpaca fibre. Since 1994, Peru has allowed the controlled export of SAC to USA, Australia, Switzerland and Canada, among others.

S. Tellez (USA): Compared to the Peruvian situation, the USA industry is very different. The population of 560,000 SAC do not come under a national breeding programme. Animals were mainly imported from Chile (200 animals), Peru (900 animals) and Bolivia (380 animals). Most alpacas belong to the Huacaya type. There are about 37,000 Suri alpacas consisting of only 12 lines with not much genetic diversity. The SAC industry is similar to Europe: animals are kept in small groups, mainly by zoos and small breeders as a hobby or second business. The acceptance of SAC meat and leather is low and the focus is on fleece products. There are two breeders' associations for llamas and alpacas where most progeny are registered and identified via DNA analysis. Some keepers also own angora goats or yaks, but there are few breeders engaged in both SAC and sheep breeding.

Common management denominators between SAC and other fibre animals

Von Thüngen (Argentine): SAC and most fibre-producing animals have in common that they are kept on natural grasslands. In the USA, however, better quality grassland than in South America is provided.

R. Morante (Peru): In South America alpacas reach 35 to 40 kg of body weight when 2 years old. A similar weight is already reached by 6 month-old alpacas in the USA. A better supply of protein and energy would improve the reproduction rate. A foaling rate of 90% could be achieved. When crias are then weaned at 6 months (25 kg body weight), there would be less mortality in females because they are not pregnant and lactating at the same time. The supplementation with lysine is crucial for fibre diameter. An increased lysine supply apparently causes an increase in fibre diameter. While Peruvian alpacas have fibre diameters of 18 μm in the first year, they may reach 30 μm in the USA. However, deficiencies in amino acids may also cause fibre quality to deteriorate. The so-called 'hunger' fibre is very fine (16-18 μm), but breaks easily during processing. Under South American conditions, no supplementation is possible, but the existing grassland could be better used, when e.g. silage is offered during periods of feed scarcity in particular during August. In January or February hay and silage could be harvested at altitudes of 3,000-4,000 m.

D. Allain (France): Under European conditions there is a large diversity of animal fibre production systems. They are mainly based on extensive grasslands with the exception of angora rabbits which are kept in stables. The use of marginal grassland may not be profitable and the added value is obtained by direct marketing. The situation in Europe is similar to the USA and Canada where the raw fleece has only a low market value compared to processed products.

G. Vila Melo (Argentina): The global market is an alternative for the commodity market. Better prices can be obtained when the fibre is classified according to fibre diameter. In Argentine for example, the llama fibre is not classified.

S. Tellez (USA): In the USA, prices were very high for live animals when only a few animals were available. Now prices have returned to more normal levels. The fibre harvested is sold via two main channels. About 20% is used in small industries for handicrafts, but there are only 4 small processing mills. The fibre-processing industry is very underdeveloped, in contrast to SAC-related research. It is estimated that about 30% of the fibre is stored or even regarded as waste, as prices for raw fibre are very low and shearing costs may exceed fibre net profits. The other option is the exportation to Peru where a well-developed fibre processing and marketing industry has been established.

R. Morante (Peru): The alpaca fibre market shares the same problems as other natural fibres. Consumers want simple and cheap clothes, this is particularly true for large markets such as China and India. Alpaca garments, however, require some special care, e.g. during cleaning. About 10% of consumers are prepared to buy fair-trade products or to pay for natural colours and sustainability of production. One major problem for South American producers is the dependency on fluctuations of the US dollar. Farmers may receive fourfold higher prices for extra fine fibre and also for better breeding animals. The market for vicuña fibre is more difficult as the dehairing of the coarser fibre is labour intensive. The goal must be to train locals to produce better quality fibre. From the world market, most of the SAC fibre is taken to Italy.

C. Renieri (Italy): Cashmere fibre is an example of a fibre on the global market with continuously high prices. Twenty years ago, Suri fibre was more important in Italy, but the fibre quality has deteriorated significantly. This problem is due to a lack of consistent selection and the fibre diameter in particular has to be improved to restore the former importance of SAC fibre.

R. Morante (Peru): The deterioration of the alpaca fibre quality can be attributed to the political situation in Peru, particularly in the 1970s and 1990s. When the first animals were exported to the USA in 1994, national interest in SAC was stimulated. For the last 5 years, the negative trend in alpaca fibre quality has stopped.

S. Tellez (USA): It is of particular interest to note that, since their importation to the USA, alpaca fibre quality has not improved and the fineness of imported animals has not been maintained.

Conclusions

- SAC and other fibre-producing animals share most management denominators.
- They are mainly kept under extensive systems with regard to nutrition and provision of stables (with the exception of the Angora rabbit). Malnutrition under these systems may reduce the quality of fibre.
- Animal fibres are subjected to the global market and receive no subsidies.
- The main traits considered in genetic selection for fibre quantity and quality are similar among the different fibre-producing species. However, the breeding indices have to consider species specific genetic correlations between traits.
- In SAC and other fibre animals there is a lack of well-organised breeding programmes and cooperation between breeders.

Recommendations for improvement

- Promotion of alternative products and niche markets for animal fibre. The present image of animal fibre could be modified towards animal fibre as a renewable raw material based on sustainable production for textile, but also for alternative purposes (e.g. insulating material).
- Increase and exchange of knowledge among scientists and breeders of SAC and other animal fibre-producing species to better withstand the global market and to support sustainable production systems.

South American camelids health

Chair: Edgar Quispe (Peru)
Chairpersons: Mª Dolores Cid Vázquez (Spain), Calogero Stelletta (Italy), Ilona Gunsser (Germany), Wilfredo Huanca (Peru)
Reporter: Edgar Quispe (Peru) and Annegret Klinkert (Germany)

Neonatal mortality is one of the most limiting factors of alpaca and llama production. Neonatal mortality implies the loss of units of production and also the loss of genetic improvement of the herd. In developed countries, neonatal mortality in alpaca production is similar to other domestic species. However, in Peru, mortalities higher than 50% or 80% of the born cria of alpaca have been reported in some years or herds. The most important risk factors for neonatal mortality are deficient hygiene conditions, overcrowding, deficient health status of mothers, poor colostrum intake, stress factors, and changes in diet. Control must be based on preventive measures which guarantee a good level of neonatal immunity and a low level of environmental contamination with infectious agents. Because there is no specific vaccination, the best strategy is good management at birth of the newborn, and good management of the females during pregnancy including vaccines.

An adequate diet for SACs is grass, good quality hay, minerals and water. However, because of the owner's desire and clever advertising by the animal food industry, many llamas and alpacas receive additional feeding stuff. This frequently causes adiposis and hyperacidity of the stomach and can lead to severe metabolic problems. The result is quite often fatty liver disease, sometimes even with fatal consequences.

Another problem is the result of absent or insufficient monitoring and treatment of parasites. The main problems may arise from parasites which need intermediate hosts living on the pasture, for example the small liver fluke (*Dicrocoelium dentriticum*). Infection by these parasites can damage the liver, resulting in cirrhosis.

Since, in most cases, the affected regions are located on the inner parts of the legs or close to the abdominal region, the first signs of mange will frequently be overlooked by the owner.

Problems with teeth root infections can be seen in animals of all ages; the reasons are unclear but it could have a genetic component. Usually, each molar tooth should have contact with the adjacent molar or premolar tooth, so no hay can stick between the teeth and irritate the gingiva. In some animals with tooth problems the reason may be the distance between these teeth.

Finally, camelids are *camelids*, not other animals. For this reason it is necessary to consider specific systems, adaptation, regions, seasons, food, pasture, and farmers, because if any of these are absent, there can be problems with SAC health.

Keyword index

A

alpaca	25, 35, 59, 65, 87, 93, 107, 111, 141, 151, 177, 209, 223
Andean	145
animal genetic resources	161
Arharmerino	51
Asip	93

B

biochemical polymorphism	97
biodiversity	161
biometrical traits	97
body measurements	209
breeding	111

C

camelid	123
– economy	195
campesino	195
carpet wool	51
cDNA polymorphism	93
CIELab Colourimetric System	79
climate	43
clustering analysis	73
coat colour	65
colour chart	79
conservation	161, 187
crossbreed	51
cytochrome b	87

D

deuterium	169
development	195
diarrhea	223
diversity	87
D-loop	87

E

enteropathogenic	223
Escherichia coli	223
evolution	107

F

farm management	209
feeding	177
feed preferences	177
fibre diameter	59, 65, 73, 107
fleece	
– classification	79
– shedding	15
– weight	65

Printed in the United States
by Baker & Taylor Publisher Services